KU-017-527

Photosynthetic Systems

Photosynthetic Systems
Structure, Function
and Assembly

SUSAN M. DANKS
Biology Tutor
Open University

E. HILARY EVANS
Department of Biology
Preston Polytechnic

and

PETER A. WHITTAKER
Department of Biology
St. Patrick's College
Maynooth, County Kildare

JOHN WILEY & SONS
Chichester · New York · Brisbane · Toronto · Singapore

Copyright © 1983 by John Wiley & Sons Ltd.

All rights reserved.

No part of this book may be reproduced by any means, nor
transmitted, nor translated into a machine language
without the written permission of the publisher.

Library of Congress Cataloging in Publication Data:
Danks, Susan M.
  Photosynthetic systems.
  Includes bibliographies and index.
  1. Photosynthesis.   2. Chloroplasts.   I. Evans,
E. Hilary.   II. Whittaker, Peter A., 1939–0000.
III. Title
QK882.D36   1983   581.1'3342   83-5831
ISBN 0 471 10250 4 (cloth)
ISBN 0 471 90178 4 (paper)

British Library Cataloguing in Publication Data:
Danks, Susan M.
  Photosynthetic systems.
  1. Photosynthesis
  I. Title   II. Evans, E. Hilary
  III. Whittaker, Peter A.
  581.1'3342   QK882
  ISBN 0 471 10250 4 (cloth)
  ISBN 0 471 90178 4 (paper)

581·13342
LIVERPOOL POLYTECHNIC
C.F. MOTT CAMPUS
LIBRARY
195904

Typesetting by Oxford Verbatim Limited
Printed by The Pitman Press, Bath, Avon.

Contents

v

Preface

Photosynthesis is arguably the most important biological activity on this planet and has rightly received a considerable degree of attention from research workers. We feel that this research has led to a situation where the basic mechanistic principles of the light-dependent processes of photosynthesis (particularly phosphorylation of ADP) have now been established and a relatively full picture of the associated light-independent metabolic processes has been achieved. Recent discoveries have also begun to illuminate the problems of chloroplast assembly and development of the photosynthetic apparatus. This book is entitled *Photosynthetic systems: structure, function and assembly* because we have tried to examine the nature of photosynthetic processes in a whole range of organisms, including bacteria. Structure is discussed in chapter 1, function in chapters 2 and 3 and assembly in chapter 4. There is one inconsistency between chapters 2 and 3 which is deliberate. In chapter 2, which deals with photosynthetic phosphorylation ('light reactions'), photosynthetic bacteria are discussed first, since a great deal of research in this field has been carried out using these organisms, which possess 1 photosystem. Chloroplasts which have 2 photosystems are dealt with subsequently. However, in our experience the dark reactions of chloroplasts are more generally taught than the metabolism of photosynthetic bacteria. Chapter 3, therefore, has a bias towards chloroplast metabolism and not so much on that in photosynthetic bacteria.

The book is written primarily for undergraduate students taking biochemistry courses, e.g., biochemistry, biology and plant physiology students. Literature references are not included in the text, but instead there is a suggested reading list at the end of each chapter. This identifies publications by many authors prominant in the specific areas of research covered in this book and also reviews from journals which students may use to open up related fields of interest. We hope that this combination will provide interesting and useful further reading for more advanced students, i.e., third year students and research workers in photosynthesis.

Throughout the text, research techniques are mentioned wherever possible. Our aim is not to give definitive descriptions of these methods, but to encourage students to make the connection between experimental techniques and accepted knowledge, which they often study separately.

The use of 'di' rather than the alternative 'bis' nomenclature was chosen, being in line with the literature searched, and to avoid confusion to the student. In two cases however, those of ribulose bisphosphate carboxylase and its substrate, 'bis' seems the more generally accepted and so is used here.

We would like to thank Drs Philip Dix, Reg England, Jacqui Manwaring, Tony Moore, Andy Morgan, Mike Tribe and Professor Mike Evans for reading and making helpful comments on various parts of the manuscript and Malcolm Danks for the design of diagrams in chapters 1 and 3 and help with the index.

We are very grateful to Dr J. W. Bradbeer, Professor J. D. Dodge, Dr S. Gibbs, Professor J. L. Hall, Professor R. Hermann, Professor J. Heslop-Harrison, Dr I. Ohad, Dr W. W. Thomson, Dr R. S. Wolfe and Dr D. R. Wolstenholme who kindly provided electron micrographs for this book.

Abbreviations

ALA	5 aminolevulinic acid
AMP, ADP and ATP	Adenosine 5′ mono-, di- and triphosphate
ATPase	Adenosine 5′ triphosphatase
Chl	chlorophyll
CO_2	carbon dioxide
CoA	coenzyme A
ctDNA	chloroplast deoxyribonucleic acid
cyt	cytochrome
DHAP	dihydroxyacetone phosphate
DCMU	3-(3,4 dichlorophenyl)-1,1-dimethylurea
DCPIP	2,6-dichloroindophenol
DBMIB	2,5-dibromo-3-methyl-6-isopropyl-p-benzoquinone
EPR	electron paramagnetic resonance
Fd	ferredoxin
FCCP	p-trifluoromethoxycarbonylcyanide phenylhydrazone
K_m	Michaelis constant
m,r or tRNA	messenger, ribosomal or transfer ribonucleic acid
MW	molecular weight in Daltons
$NAD^+/NADH$	Nicotinamide adenine dinucleotide (oxidized/reduced)
$NADP^+/NADPH$	Nicotinamide adenine dinucleotide phosphate (oxidized/reduced)
PC	plastocyanin
PEP	phosphoenol pyruvate
Pi	inorganic phosphate
PQ/PQH_2	Plastoquinone (oxidized/reduced)
RBPC	ribulose bisphosphate carboxylase
S	Svedberg constant
SDS	sodium dodecyl sulphate

1
Introduction

1.1 Introduction to prokaryotic and eukaryotic cells

Living organisms can be divided into two groups, the prokaryotes and the eukaryotes. These two groups differ fundamentally in the structure and biochemistry of their cells (see table 1.1).

Prokaryotic cells (bacteria) generally have a cell wall and cell membrane enclosing the cytoplasm, but no internal cell membranes. The prokaryotic cell's DNA is not contained within a nuclear membrane, but may be attached to the cell membrane. Electron transfer reactions associated with ATP synthesis occur within the cell membrane. The ribosomes, which are involved in protein synthesis, are mostly suspended in the cytoplasm but may be attached to the cell membrane.

Eukaryotic cells, e.g., plant and animal cells, contain complex internal cell membranes which surround subcellular structures called organelles. The DNA is enclosed in the nucleus in the form of chromosomes. Electron transfer reactions occur primarily in the inner membranes of chloroplasts and mitochondria. Ribosomes, which have different sedimentation characteristics from those of prokaryotes, may be attached to the endoplasmic reticulum or in the cytosol (the soluble part of the cytoplasm). The other organelles present in eukaryotic cells are lysosomes, Golgi apparatus and vacuoles.

Photosynthesis is the process whereby light energy from the sun is converted to chemical energy and conserved in the form of ATP and NAD(P)H, which can be used to drive the biosynthesis of organic molecules such as glucose and amino acids. Photosynthesis can occur in both prokaryotes and eukaryotes, e.g., photosynthetic bacteria and green plants. (For convenience these are referred to collectively in this book as photosynthetic systems.)

1.1.1 Classification

This section gives a brief classification of prokaryotic and eukaryotic photosynthetic organisms. It is widely thought that photosynthesis is carried out only by eukaryotic plant cells, but about half of all photosynthetic reactions occur in prokaryotes.

A classification of prokaryotic photosynthetic organisms is shown in figure

1

Table 1.1 The major differences between prokaryotic and eukaryotic cells

	Eukaryotic cells	Prokaryotic cells
	Cell components divided between subcellular structures called organelles, e.g., nucleus, mitochondria, lysosomes,[1] Golgi apparatus[2] and endoplasmic reticulum. Plant cells also contain vacuoles[3] and plastids including chloroplasts.	Cell components contained within the cell membrane, no internal membrane systems, i.e., no organelles.
Cell membrane/ cell wall	Selectively permeable to ions. Contains about 50% protein and 50% lipid. Some animal cell membranes have a cell coat on the outer side, composed mainly of polysaccharide. Plant cells have a rigid cell wall consisting of cellulose polysaccharides and protein.	Selectively permeable to ions. Contains about 55% protein and 45% lipid – sometimes invaginated. Surrounded by a cell wall of rigid polysaccharide cross-linked with peptide chains which protects the cell from hypotonic solutions, i.e., swelling.
DNA	Surrounded by a membrane forming the nucleus. Combined with histone proteins and forming chromosomes. Mitochondria and chloroplasts contain some circular DNA.	Tightly coiled but not surrounded by a membrane, may be attached to the cell membrane.
Electron transport (ATP formation)	Occurs in mitochondria and chloroplasts. The proteins of the electron transport system are part of the inner mitochondrial membrane or thylakoid membranes in chloroplasts.	Proteins of the electron transport system are associated with the cell membrane. Chromatophores in photosynthetic bacteria are invaginations of the cell membrane.
Ribosomes	Attach to endoplasmic reticulum membrane system. Composed of 60S and 40S components.	Attach to cell membrane. Composed of 50S and 30S components.

[1] Lysosomes contain hydrolytic enzymes.
[2] Golgi apparatus is involved in secretion.
[3] Vacuoles act as stores of sugars, salts etc.

1.1, emphasizing the pigments present in them. Three groups of photosynthetic prokaryotes are known. Bacteria which have only a single photosystem, and use a reductant other than water (e.g., H_2S, reduced organic compounds or H_2) to provide the reducing equivalents for photosynthesis, are grouped into the Rhodospirillales.

The second major group of photosynthetic prokaryotes are the cyanobacteria, formerly known as Cyanophyta or blue-green algae. They have two photosystems, use water as a reductant and like plants and algae, evolve oxygen in the light. Recently, however, it has been decided to classify them as bacteria because of their prokaryotic characteristics. Their classification is complex and based mainly on morphology rather than pigmentation, which is the main feature used for classifying other photosynthetic microorganisms. Unfortunately, as further studies of cyanobacteria are revealing, morphological classification gives no guidance about metabolic complexity. Michael Herdman has recently identified three different sized genomes in cyanobacteria, 2a, 4a and 6a, compared with 2a, the genome size of the average bacterium. The genome size appears to be related to morphological and/or metabolic complexity and may provide the future basis of a more coherent classification. The majority of biochemical studies on cyanobacteria have used organisms of the following species: *Synechoccus, Phormidium, Anacystis, Aphanocapsa* (unicells); *Anabena* (filamentous); *Chlorogloepsis, Nostoc* (filamentous, aseriate mixed morphology).

The third group of photosynthetic prokaryotes, the Prochlorophyta, to date has only a single species called *Prochloron*, which contains chlorophyll *a* and *b* but lacks phycobilins. It is thought to be an evolutionary precursor of chloroplasts.

In recent years it has been found that a group of salt-tolerant bacteria, Halobacteria, are capable of ATP synthesis coupled to a light-driven reaction, totally unrelated to the photosynthetic electron transport chains described here and in chapter 2. Some more details of this unusual reaction will be given in section 2.13.

There is much dispute over the details of algal classification. In general, they are eukaryotic cells containing chloroplasts and some are classified as plants. However, like photosynthetic prokaryotes, their classification is based on the following characteristics:

(1) Pigments: their nature or chemical composition.
(2) Reserve food products or assimilatory products of photosynthesis.
(3) Flagella.
(4) Cell walls.
(5) Morphological characteristics of cells and thalli.
(6) Life history and reproduction.

To try to illustrate the variations in algae, table 1.2 shows some of the differences between six algal Divisions. All six have an oxygen-evolving, 2-photosystem photosynthetic apparatus, but differ markedly in their ultrastructure

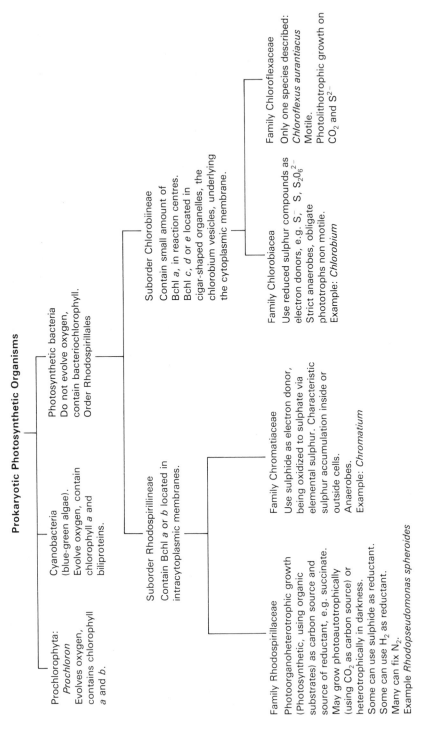

Prokaryotic Photosynthetic Organisms

Prochlorophyta:
Prochloron
Evolves oxygen,
contains chlorophyll
a and *b*.

Cyanobacteria
(blue-green algae).
Evolve oxygen, contain
chlorophyll *a* and
biliproteins.

Photosynthetic bacteria
Do not evolve oxygen,
contain bacteriochlorophyll.
Order Rhodospirillales

Suborder Rhodospirillineae
Contain Bchl *a* or *b* located in
intracytoplasmic membranes.

Suborder Chlorobiineae
Contain small amount of
Bchl *a*, in reaction centres.
Bchl *c*, *d* or *e* located in
cigar-shaped organelles, the
chlorobium vesicles, underlying
the cytoplasmic membrane.

Family Rhodospirillaceae
Photoorganoheterotrophic growth
(Photosynthetic, using organic
substrates) as carbon source and
source of reductant, e.g. succinate.
May grow photoautotrophically
(using CO_2 as carbon source) or
heterotrophically in darkness.
Some can use sulphide as reductant.
Some can use H_2 as reductant.
Many can fix N_2.
Example *Rhodopseudomonas spheroides*

Family Chromatiaceae
Use sulphide as electron donor,
being oxidized to sulphate via
elemental sulphur. Characteristic
sulphur accumulation inside or
outside cells.
Anaerobes.
Example: *Chromatium*

Family Chlorobiacea
Use reduced sulphur compounds as
electron donors, e.g. S^-, S, $S_2O_6^{2-}$
Strict anaerobes, obligate
phototrophs non motile.
Example: *Chlorobium*

Family Chloroflexaceae
Only one species described:
Chloroflexus aurantiacus
Motile.
Photolithotrophic growth on
CO_2 and S^{2-}

Figure 1.1   Classification of prokaryotic photosynthetic organisms

Table 1.2 Differences between six algal Divisions

Division	Chrysophyta (golden algae) (diatoms)	Pyrrophyta (Dinoflagellates)	Euglenophyta (euglenoids)	Chlorophyta (green algae)	Phaeophyta (brown algae)	Rhodophyta (red algae)
Chlorophyll a	+	+	+	+	+	+
Chlorophyll b	–	–	+	+	–	–
Chlorophyll c	+	+	–	–	+	–
Chlorophyll d	–	–	–	–	–	sometimes
Carotenoids						
Carotenes	+	+	+	+	+	+
Fucoxanthin	+	+	–	–	+	–
Peridinium	–	+	–	–	+	–
Others	–	–	–	lycopene, lutein	–	zeaxanthin
Phycobilins						
Phycocyanin	–	+	–	–	–	+
Phycoerythrin	–	+	–	–	–	+
Storage material	chrysolaminarin, oils	starch, oils	paramylon, oils	starch	laminarin, oils mannitol	floridean starch, oils
Flagella	1 or 2	2 lateral 1 trailing 1 girding	1, 2 or 3 equal	1, 2, 4 to many	2	none
Notes			has a gullet			
Number of thylakoids grouped together in chloroplast	3	3	3	2 or more	3	1
Example	Ochromonas (Chromulina)	Amphidinium	Euglena	Chlamydomonas		Prophyridium

and the pigmentation of their chloroplasts. Table 1.2 shows the major differences in pigmentation, storage material, thylakoid grouping and flagella.

Another group of eukaryotic photosynthetic organisms are the higher orders of plants: mosses, liverworts, ferns, gymnosperms and flowering plants. They all contain chloroplasts which have a similar pigment composition to the Chlorophyta (table 1.2), i.e., they contain chlorophyll a and b and some other pigments such as carotenoids. Their chloroplasts carry out oxygen-evolving, 2-photosystem photosynthesis.

1.2 Structure and ultrastructure

This section does not pretend to cover the structure of photosynthetic organisms in great detail, but tries to show the main features that enable them to carry out their unique reactions. The main structures shown here are therefore chloroplasts from eukaryotic cells and chromatophores in prokaryotic cells.

To observe the detailed structure of such small particles as chloroplasts, which are invisible to the naked eye and many of whose components are impossible to resolve under optical microscopes, the electron microscope is used. Essentially this works on the same principle as the light microscope, except that a beam of electrons, instead of light, 'illuminates' the sample, and a series of electric coils act as electron lenses bending the electron beam in the same manner that optical lenses bend a light beam.

Two methods of preparing samples for the electron microscope are widely used. The more common, as in optical microscopy, is to fix and stain the material under investigation. Fixing, a measure intended to preserve the ultrastructural detail in its natural state, is done by immersion in an aqueous solution of osmium tetroxide (OsO_4) or potassium permanganate ($KMnO_4$) or in glutaraldehyde. The sample is then dehydrated using ethanol and then embedded in a synthetic resin such as Araldite. Then the sample is sliced to obtain a specimen for the electron microscope of 50–100 nm thickness. Solutions of lead hydroxide, lead citrate or uranyl acetate can then be used to stain the specimen if necessary.

The second technique, particularly useful for studying membrane ultrastructure, is called freeze-etching. The sample is rapidly frozen to a temperature near that of liquid nitrogen ($-195°C$) and then kept cold in a vacuum chamber. Slicing the sample under these conditions exposes regions from which ice sublimes, leaving membrane edges clearly exposed. A replica is then made by shadowing with platinum and carbon. It is the replica that is observed under the electron microscope after the remains of the sample have been removed.

In green plants and photosynthetic algae, photosynthesis occurs in subcellular organelles called chloroplasts. Chloroplasts are one of a group of organelles called plastids, all of which comprise a double membrane surrounding an internal membrane system. Some plastids act mainly as storage organelles, e.g., amyloplasts contain starch and elaioplasts contain lipids. Chloroplasts are the most complex of the plastids and are thought to develop from precursors called proplastids, which are 1–3 μm long with only a few internal membranes. Chloroplast development from proplastids involves an increase in volume of the plastid, rapid chlorophyll biosynthesis and an increase in internal membranes. This development will be discussed in more detail in chapter 4.

Figure 1.2 shows an electron micrograph of a chloroplast in the cell of the halophyte *Suaeda maritima* (which grows in salt marshes): the sample has been

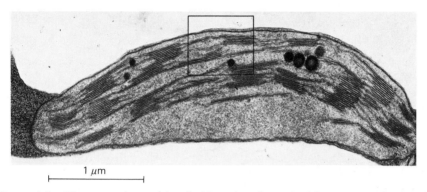

1 μm

Figure 1.2   Electron micrograph of chloroplast from *Suaeda maritina*, fixed and stained. (Courtesy of Professor J. L. Hall)

fixed and stained. Figure 1.3 is a schematic representation of part of the same chloroplast enlarged two-fold to show all the features. Chloroplasts are generally ellipsoidal in shape and vary in length from 4 to 10 μm. There can be 1 to 100 chloroplasts per cell. The chloroplast in figure 1.2 is surrounded by a thin layer of cytosol and a vacuole can also be seen. The chloroplast has a double boundary membrane referred to as the chloroplast envelope. Each of these two membranes is 6–8 nm thick and a gap of 10–20 nm exists between them.

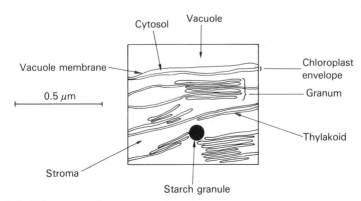

Figure 1.3   Diagrammatic representation of the structure contained within the square drawn on figure 1.2, magnified two-fold

8

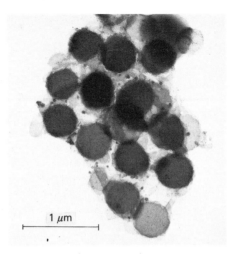

Figure 1.4 Electron micrograph of a segment of the internal membrane system of chloroplasts from Swiss Chard, fixed and stained, showing various stages of thylakoid growth. (Reproduced from J. Heslop-Harrison, *Science Progress*, 1966, **54**, 538, 539 by permission of Blackwell Scientific)

Chloroplast membrane structure will be discussed in section 1.2.1. Inside the envelope is the liquid part of the chloroplast, the stroma, which appears granular under the electron microscope and is rich in enzymes. Within the stroma is an inner membrane system consisting of flattened sacs or discs called thylakoids. These can often be seen arranged in stacks known as grana, which are linked by hollow membranous bridges, the lamellae (sometimes referred to as frets or stroma thylakoids). The whole system of interconnected chambers is believed to enclose a single continuous space, the thylakoid space. Figure 1.4 shows an isolated segment of a thylakoid membrane system. Various stages of thylakoid growth can be seen: small, presumably immature granum thylakoids appear superimposed upon fully extended ones. Figure 1.5 gives a representa-

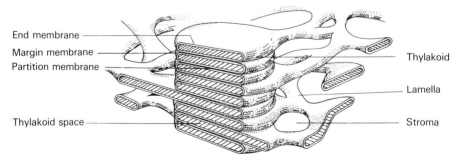

Figure 1.5 Diagrammatic representation of thylakoids, grana and lamellae, indicating how the space within them may be continuous.

tion of how the grana, thylakoids and lamellae are believed to be linked so as to enclose a single continuous space.

Figure 1.6 shows part of a chloroplast from a green tomato fruit prepared by fixing and staining. Grana can be seen clearly linked by lamellae. Part of the chloroplast envelope is visible and also two plastoglobuli, which are believed to be sites of lipid storage. This electron micrograph clearly shows that membranes in different parts of the grana have different surroundings. Some, the so-called end membranes and margins, are in direct contact with the stroma, while others, the so-called partition membranes, are not. The significance of this is discussed in section 1.5.2 (figure 1.22).

Figure 1.7 shows chloroplasts of *Amphidinium britanicum*, from the algal class Dinophyceae (Pyrrophyta in table 1.2). Most algal Divisions, with the exception of Chlorophyta (green algae), do not have large grana. In figure 1.7

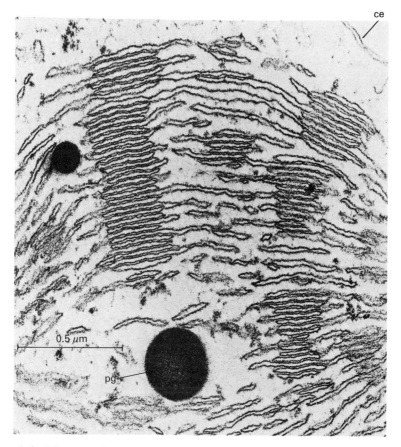

Figure 1.6   Electron micrograph of part of a chloroplast from green tomato fruit fixed and stained. pg: plastoglobulus, ce: chloroplast envelope. (Copyright © 1974 McGraw-Hill Book Co (UK) Ltd. From Chapter 4 by Professor W. W. Thompson in Robards: *Dynamic Aspects of Plant Ultrastructure*. Reproduced by permission of the publisher)

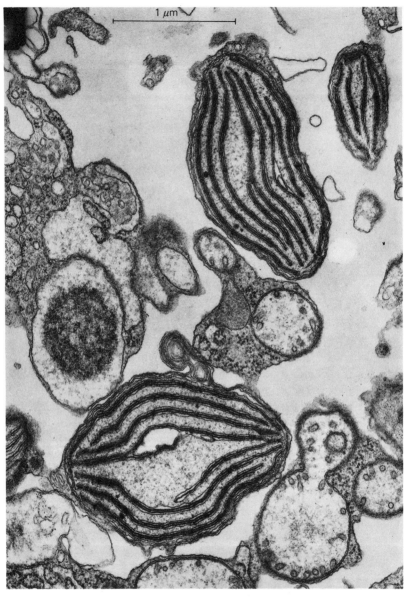

Figure 1.7 Electron micrograph of chloroplasts from the alga *Amphidinium. britanicum*, fixed and stained (Courtesy of Professor J. D. Dodge) (With permission from J. D. Dodge, *The Fine Structure of Algal Cells*, 1973. Copyright: Academic Press Inc. (London) Ltd.)

the thylakoids can be seen arranged predominantly in threes, a very common characteristic of algal chloroplasts. However, other arrangements of thylakoids do exist in algae: Crytophyceae generally have the thylakoids stacked in pairs and Rhodophyceae have the simplest arrangement with just single thylakoids in the stroma. Green algal chloroplasts have grana similar to those in plant chloroplasts. Lipid globules are often seen in the electron micrographs of algal chloroplasts, or starch granules if starch is the main storage material.

Some plant chloroplasts have no grana; figure 1.8 shows a bundle sheath cell chloroplast from *Zea mays* (maize), a C_4 plant showing reduced numbers and smaller sized grana. Studies on mature C_4 plants, chlorophyll-deficient mutants

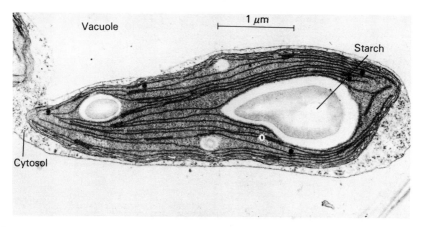

Figure 1.8   Electron micrograph of a chloroplast from bundle sheath cells of *Zea mays L.*, fixed and stained. (Copyright © 1974 McGraw-Hill Book Co (UK) Ltd. From Chapter 4 by Professor W. W. Thompson in Robards: *Dynamic Aspects of Plant Ultrastructure*. Reproduced by permission of the publisher)

and algae with only a few grana suggest that the occurrence of certain membrane proteins or pigment-protein complexes may be related to the degree of thylakoid stacking to form grana (section 1.5.2). The bundle sheath chloroplast in figure 1.8 contains large starch grains each surrounded by a clear zone. These zones may well be artefacts caused by shrinkage during fixation and dehydration.

Photosynthetic bacteria are protected by a cell wall which surrounds the cell membrane. Invaginations of the cell membrane form structures called chromatophores, which are the site of the electron transport reactions. Figure 1.9a shows isolated *Rhodospirillum rubrum* (*Rs. rubrum*) cells, which have been fixed and stained, clearly showing all these features of photosynthetic bacteria. Figure 1.9b shows isolated chromatophores from *Rs. rubrum* which have been freeze-etched. Figure 1.10 shows a *Rhodopseudeomonas capsulata* (*Rp. capsulata*) cell which has been freeze-etched. The cell wall, cell membrane and chromatophores can all be seen and this technique also shows the proteins

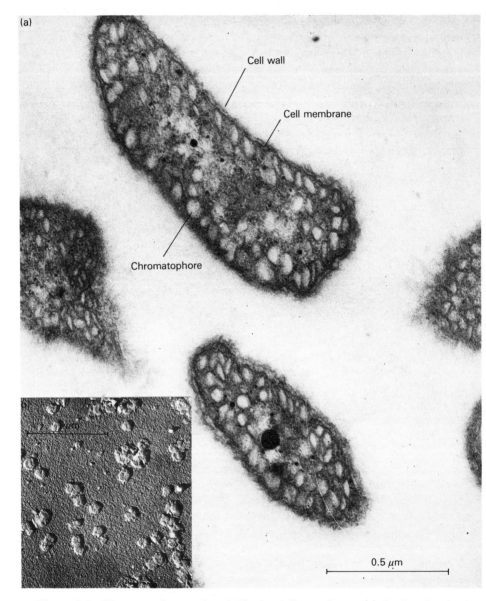

Figure 1.9 Electron micrographs of *Rhodospirillum rubrum*: (a) fixed and stained cells, (b) isolated freeze-etched chromatophores. (Reproduced from Vatter and Wolfe, *J. Bacteriol.*, 1958, **75**, 480–488 by permission of the American Society for Microbiology)

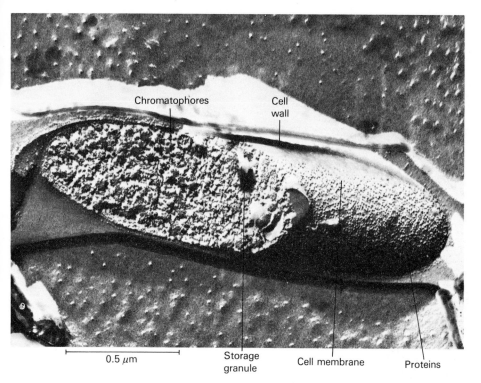

Chromatophores
Cell wall

0.5 μm
Storage granule
Cell membrane
Proteins

Figure 1.10   Electron micrograph of a freeze-etched cell from *Rhodopseudomonas capsulata*, cell wall, cell membrane, membrane proteins, chromatophores, storage granules

embedded in the cell membrane. In figure 1.11, *Rps. molischianum*, which has been fixed and stained, the cell wall and cell membrane can be seen clearly. This sample was taken during a study of the effects of different light conditions and temperatures on the internal membranes (chromatophores) of this photosynthetic bacterium. The bacteria were incubated in medium at 34°C with light of 65000 lx shining on them. This apparently caused the membrane to form thylakoid-like discs, a stack of which (three) can clearly be seen in figure 1.11. It seems that the precise structure of the membranes' invaginations in photosynthetic bacteria may also depend, like the stacking of thylakoids, on the presence or absence of certain proteins or pigments since changes in light conditions can also alter the amount of pigment present.

Figure 1.12 shows a freeze-etched cell of the cyanobacterium (blue-green alga) *Chlorogloepsis fritschii* (*Chlorogloea fritschii*). Close to the cell wall and cell membrane a storage granule and many intracellular membranes or lamellae can be seen. These membranes are the site of the electron transport reactions in this species. Unlike the chromatophores of the photosynthetic bacterium *R. rubrum* (figure 1.9), which take the form of discrete invaginations of the

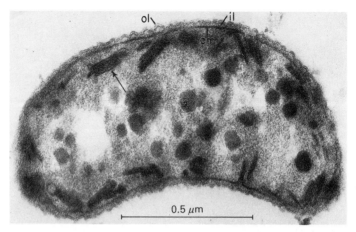

Figure 1.11   Electron micrograph of *Rp. molischianum*, fixed and stained, ol and il: outer and inner layer of cell wall, cm: cell membrane; arrow indicates unusual structures (discs) in cytoplasm. (Reproduced from S. P. Gibbs *et al.*, *The Journal of Cell Biology*, 1965, **26**, 395–412 by copyright permission of The Rockefeller University Press)

bacterial cell membrane, the lamellar system of *Chloroeglopsis fritschii* cannot be isolated as membrane vesicles.

1.2.1   The composition of chloroplast membranes

The generally accepted model for membrane structure is the fluid mosaic model (figure 1.13), which has been refined by Singer and Nicolson. In this model proteins are able to move in a fluid bilayer of lipid. The proteins may be intrinsic or extrinsic. Intrinsic proteins are embedded within the lipid bilayer, but they may be more on one side of the membrane than the other or they may extend right through the membrane. In the chloroplast thylakoid membrane plastocyanin is on the inner side of the thylakoid membrane while cytochrome b_6 is on the stroma side of the membrane. Proteins that stretch right through the membrane are the ion translocators (section 3.9), they tend to be mainly hydrophobic because of their interaction with the lipids. It is difficult to isolate them active and free from lipid, since they tend to form hydrophobic aggregates which often precipitate during the preparation procedure.

Extrinsic proteins, which interact with the polar head groups in the lipid bilayer, are not generally part of the membrane and can be fairly easily removed; for example, ferredoxin reductase and cytochrome f can be removed from the thylakoid membrane and cytochrome c can be removed from the mitochondrial membrane.

It is often difficult to ascertain experimentally where in a membrane a particular protein is located. A technique that has proved useful involves treating intact organelles or vesicles with reagents which cannot cross the membrane, so that only the exposed, outermost proteins are modified by the

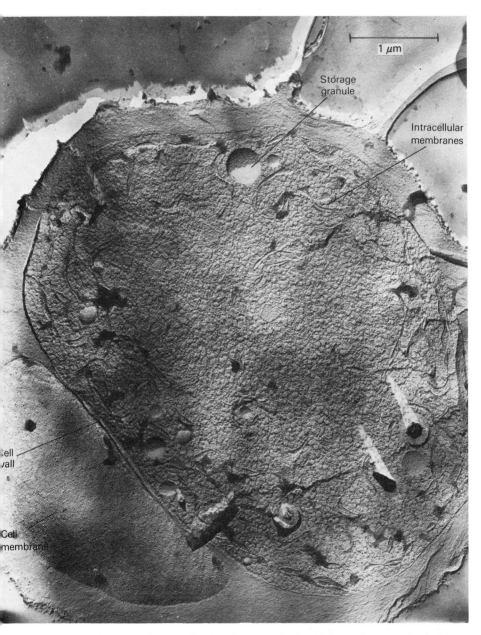

1 µm

Storage
granule

Intracellular
membranes

:ell
vall

Cell
membrane

Figure 1.12   Electron micrograph of a freeze-etched cell from the cyanobacteria
Chlorogloea fritschii, cell wall, cell membrane, intracellular membranes, storage granule

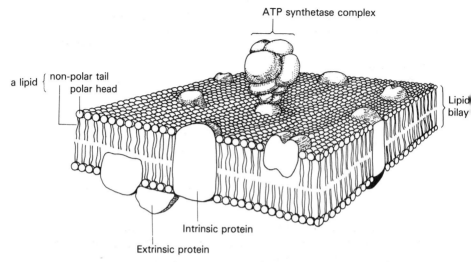

Figure 1.13   Diagrammatic representation of the fluid mosaic model of membrane structure showing intrinsic and extrinsic proteins and the ATP synthetase complex as well as the lipid bilayer

reagent. The proteins can be treated by iodination using lactoperoxidase, with diazo benzene sulphonate (DABS) or with their specific antibody. Figure 2.38 shows the possible spatial arrangement of the components of the chloroplast electron transport chain in the thylakoid membrane.

By various techniques including chromatography, the types and amounts of proteins and lipids in membranes can be measured. The chloroplast envelope membranes contain 30 % protein, including an ATPase, nitrate reductase and ion translocators. The remaining 70 % is lipid material. By comparison the protein:lipid ratio found in the lamellar and thylakoid membrane is about 60:40. Here, as in the inner mitochondrial membranes, where the protein: lipid ratio is as much as 75:25 and similar functions are carried out, proteins are abundant and enzymatic activity is high. The functions of these proteins, which are mainly electron transfer proteins and ion translocators, are further discussed in chapters 2 and 3.

The types of lipid present in chloroplast membranes are very different from those in mitochondria, which contain about 80 % phospholipid, including the unusual lipid cardiolipin which seems to confer chloride impermeability on the inner mitochondrial membrane. The thylakoid and lamellae membranes in chloroplasts contain only 10 % phospholipid (see table 1.3) and this is nearly all phosphatidyl glycerol with only traces of other phospholipids. There is no cardiolipin, and the membrane is freely permeable to chloride ions (Cl^-). The majority of the lipids present are glycolipids, i.e., monogalactosyl diacylglycerol (MGDG), digalactosyl diacylglycerol (DGDG) and sulpholipid (figure 1.14). The pigments, e.g., chlorophyll, make up the remaining 20–25 % of the lipid content of these membranes. The fatty acids attached to these

Table 1.3   Lipid content of chloroplast membranes

	Approx. % of total lipid
Phospholipids	
Phosphatidyl glycerol	10
Glycolipids	
Monogalactosyl diacylglycerol (MGDG)	40
Digalactosyl diacylglycerol (DGDG)	20
Sulpholipid	10–15
Pigments	
Chlorophylls and others	20–25

thylakoid lipids are mainly the unsaturated fatty acid linolenic acid (18:3) although some palmitic acid (16:0) is also present. The high degree of unsaturation causes the thylakoid bilayer to be very fluid.

The inner membranes of photosynthetic bacteria (cell membranes) have a very similar composition to the inner mitochondrial membrane, the proteins in these membranes being mainly associated with electron transfer reactions.

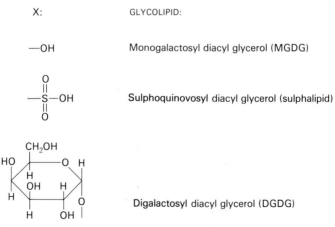

Figure 1.14   Structure of the major lipids in the chloroplast membrane

1.3 Nutrition

All living organisms utilize organic molecules such as glucose and amino acids for growth and reproduction, since these are the precursors of carbohydrates, lipids, proteins and nucleic acids. However, organisms can be divided into two groups on the basis of their source of carbon for these precursor molecules, i.e., autotrophic (self-feeding) and heterotrophic (other feeding) cells. Autotrophic cells can utilize carbon directly from the inorganic molecule CO_2, but heterotrophic cells require carbon in an organic form, e.g., glucose. These two groups can also be divided according to what source of energy their metabolism employs; this could be light or reduction–oxidation (redox) reactions like those occurring in the tricarboxylic acid cycle (TCA cycle). Table 1.4 shows these subdivisions and examples of organisms from each group. This book will concentrate on the light-harvesting systems unique to phototrophs, i.e., the photoautotrophs and photoheterotrophs (sometimes referred to as photolithotrophs and photoorganotrophs). The organisms which utilize light are the photoautotrophic plants and cyanobacteria and the photoheterotrophic bacteria, e.g., Rhodospirillaceae.

Organisms do not always fall neatly into the subdivisions in table 1.4. For instance, higher plants become heterotrophic in the dark, utilizing carbon from

Table 1.4   Classification according to carbon and energy sources

	Autotrophs	Heterotrophs
Carbon source Energy source	CO_2	Organic compounds
Light	Photoautotrophs, e.g., plants and cyanobacteria	Photoheterotrophs, e.g., *Rhodospirillaceae*
Redox	Chemoautotrophs, e.g., *Nitrobacter, Hydrogenomonas*	Chemoheterotrophs e.g., higher animals

the organic molecules produced in the light and energy from redox reactions. Organisms can also be classified according to whether their ultimate electron acceptor is oxygen (aerobic organisms) or organic/inorganic molecules (anaerobic organisms). A third group, the facultative organisms, e.g., *Rp. capsulata* and *Rp. spheroides*, can live anaerobically or aerobically, as conditions dictate, although if oxygen is available they will generally utilize it because this is more economical. Organisms which can only live anaerobically are called obligate anaerobes, for example, *Chlorobium*. Such organisms often possess some means of removing oxygen from their environment, e.g., leghaemoglobin in the nitrogen fixing organism *Rhizobium* (section 3.8.1).

Autotrophic and heterotrophic organisms are metabolically interdependent; figure 1.15 shows how carbon is cycled between the two groups of organisms.

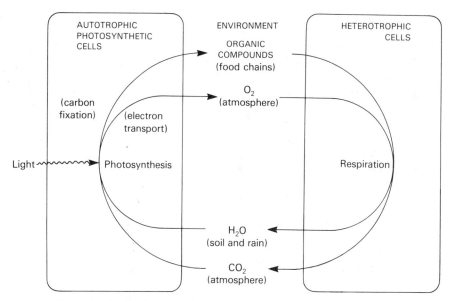

Figure 1.15   Cycling of carbon between heterotrophs and autotrophs

Autotrophs utilize CO_2 to build up organic molecules which can then be further utilized or broken down by heterotrophs, ultimately producing CO_2 once more. It is the TCA cycle which generates most of the CO_2 produced by higher animals. A similar scheme can be drawn to show how nitrogen is also cycled between autotrophic and heterotrophic organisms (figure 1.16).

The importance of light to all living organisms may be summarized in the following way: green plants use light from the sun to biosynthesize organic molecules from inorganic molecules; animals then eat either plants or animals that have eaten plants. The sources of the energy that we consume in our

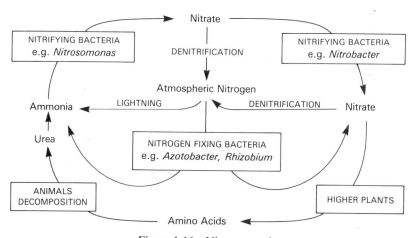

Figure 1.16   Nitrogen cycle

homes, factories and cars, for example (coal, wood and oil) is all derived from plant material. Hence light energy from the sun is the ultimate source of energy for living organisms.

Using carbohydrate, e.g., a polysaccharide, as an example of a food macro-molecule taken in by animals, figure 1.17 shows schematically what could happen to this food in an animal cell. The carbohydrate is broken down in the digestive tract into its monomers, mainly glucose. The glucose can be trans-

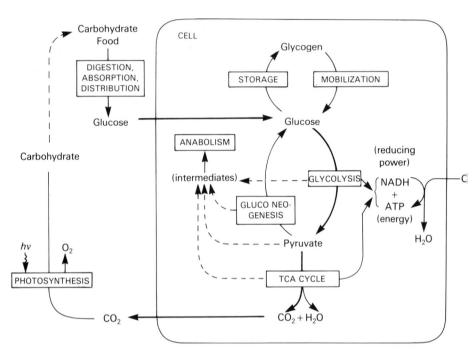

Figure 1.17   Scheme showing some possible fates of carbohydrate in a heterotrophic cell

ported into the cell where it could be stored as carbohydrate (in the liver) in times of plenty, or broken down via glycolysis and the TCA cycle to carbon dioxide and water. This oxidation of glucose gives rise to reducing power (NAD(P)H) and energy (ATP), which can be used for other biosynthetic reactions in the cell, e.g., gluconeogenesis and protein synthesis. There is a net loss of carbohydrate from the cell if the glucose is completely oxidized to CO_2 and H_2O, and since the animal cannot biosynthesize glucose from CO_2 and H_2O, the carbohydrate store has to be replenished by consuming more plant-derived food material (organic molecules).

Chapter 2 discusses how light energy is converted to chemical energy (in the form of ATP) and reducing power (reducing equivalents) in the form of NAD(P)H, the 'light reactions'. Chapter 3 discusses how the ATP and NAD(P)H is used to produce organic molecules, the 'dark reactions'.

In the study of living organisms, ATP (adenosine 5' triphosphate) or its hydrolysis is often referred to as providing the energy which drives biosynthetic reactions. The structure of ATP is shown in figure 1.18a. Although its structure includes adenine, a purine also found in nucleic acids and NAD(P)H, its major function as an immediate donor of free energy depends to a great extent on the triphosphate structure. During metabolic reactions such as cell wall

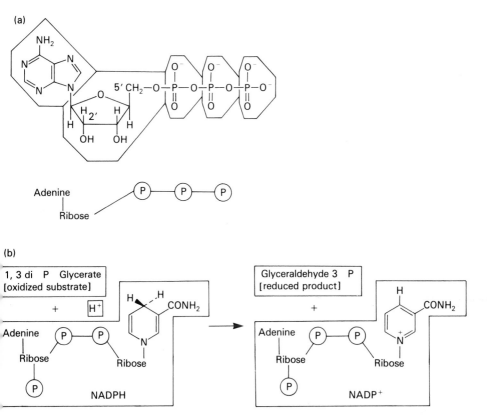

Figure 1.18 (a) Structure of ATP. (b) Structure of NADPH showing its involvement in the reaction catalysed by glyceraldehyde 3 phosphate dehydrogenase. The structure of NADH is the same as that of NADPH without the phosphate attached to the ribose ring at the 2' position

biosynthesis, the movement of flagella, muscle contraction and gluconeo-genesis, ATP is converted to adenosine 5' di-phosphate (ADP), but this reaction is not simply the hydrolysis of a phosphate group from ATP, which is an exergonic reaction:

$$ATP + H_2O \rightleftharpoons ADP + Pi \qquad \Delta G^0 = -30.5 \text{ kJmol}^{-1}$$

The energy liberated by this hydrolysis, instead of escaping, is used by linking it to an energonic reaction, i.e., one that requires energy to go in the preferred direction. For example, the reaction catalysed by pyruvate carboxylase is able to produce oxaloacetate because of the involvement of ATP.

Pyruvate $+ CO_2 + ATP \rightleftharpoons$ Oxaloacetate $+ ADP + Pi \quad \Delta G^0 = -2.1\,kJmol^{-1}$

In fact ATP usually participates in such reactions by acting as a phosphate donor after the enzyme is involved; the phosphorylated enzyme being a reaction intermediate. NADH and NADPH play a similar role in metabolism. Figure 1.18b shows the structure of NADPH; that of NADH is the same except that it does not have a phosphate group attached to the ribose ring at position 2'. NADH is produced by the electron transfer reactions of photosynthetic bacteria and by the TCA cycle, while NADPH is produced by the electron transfer reactions in chloroplasts. Figure 1.18b shows a reaction from the Calvin cycle which uses NADPH. The enzymes which use NAD(P)H as coenzymes are called pyridine-nucleotide-linked dehydrogenases. Malate dehydrogenase is another example of one of these enzymes, catalysing the following reaction:

$$NADH + H^+ + Oxaloacetate \rightleftharpoons NAD^+ + Malate$$

Although some ATP is made by substrate-level phosphorylations, e.g., that catalysed by pyruvate kinase in glycolysis, the bulk of ATP required by cells is made by phosphorylation of ADP accompanying or coupled to electron transport. In photosynthetic systems this is driven by light. Inevitably throughout this book there will be comparisons between photosynthetic and aerobic oxidation systems; the latter include mitochondria and aerobic bacteria, since ATP in mitochondria and aerobic bacteria is formed by the phosphorylation of ADP which accompanies mitochondrial or bacterial electron transport. In mitochondria, however, this electron transport is driven by reduced pyridine-nucleotides, which are formed during the oxidation of metabolites (SH_2) in the TCA cycle (redox reactions).

$$NAD^+ + SH_2 \rightleftharpoons NADH + H^+ + S$$

1.4 Light

For photosynthetic systems to carry out their primary biosynthesis, light has to be absorbed and converted to chemical energy. This will be discussed in detail in chapter 2, but first, what is light? Light is the visible part of the electromagnetic spectrum. Its wavelength ranges from 380 nm (violet) to 780 nm (red). The electromagnetic spectrum is shown in figure 1.19. Radiations of wavelengths below 380 nm are called ultraviolet and those above 780 nm are called infrared. Generally these are not useful to photosynthetic systems, although there are some bacteria which can use far-red up to 1000 nm. Energy is radiated in discrete packets or quanta; in the case of light these are called photons. The energy of a single quantum of radiation (E) is defined as:–

$$E = h\nu \tag{1}$$

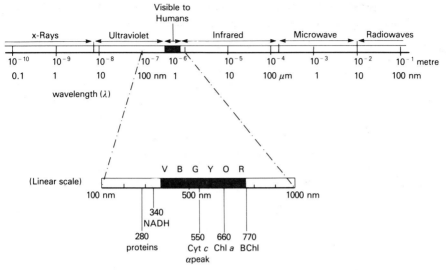

Figure 1.19   Electromagnetic spectrum

where h is Planck's constant (6.625×10^{-34} J) and ν (nu) is the frequency of the radiation, i.e., the number of complete waves per second. On some diagrams the action of light is represented by $h\nu \rightarrow$. Nu (ν) is inversely proportional to the wavelength (λ (lambda), in metres) of the radiation, i.e., $\nu \propto 1/\lambda$

$$\text{or } \nu = \frac{c}{\lambda} \tag{2}$$

where the constant of proportionality (c) is the speed of light (2.998×10^8 m s^{-1}). Therefore, combining (1) and (2)

$$E = \frac{hc}{\lambda}$$

which means that the energy of a photon is inversely proportional to the wavelength of the light. So blue light $\lambda = 500$ nm is of greater energy than red light $\lambda = 700$ nm (E is respectively 23.9 and 17.1 joules per mole of photons).

For the light energy to be useful to the photosynthetic systems it must first be trapped or absorbed by pigments. Each pigment is capable of absorbing photons at certain characteristic wavelengths. A variety of such pigments is usual in photosynthetic systems, enabling them to collect light of various wavelengths. The absorption spectrum of the pigments indicates which wavelengths can be harvested initially.

The most important type of pigment found in photosynthetic systems is chlorophyll, but there are others, e.g., carotenoids (section 2.2). Pigment molecules are believed to occur in pigment–protein complexes within the chloroplast thylakoid membrane (section 1.5.2). The energy initially absorbed is 'funnelled' to a particular chlorophyll molecule called the reaction centre chlorophyll. The pigment–protein complex which includes the reaction centre is called a photosystem. Plant and algal chloroplasts have been found to contain two types of photosystem, called photosystem 1 and 2. In these two photosystems the reaction centres are called P700 and P680, respectively, where P stands for pigment and 700 and 680 for their absorption maxima in nm. The membranes of photosynthetic bacteria contain only one type of photosystem and the reaction centre is also defined by its absorption maximum, e.g., P870 or P890.

All the pigments found in photosynthetic systems possess key structures called chromophores, where the energy from the light is absorbed. When this occurs the chromophore electrons are caused to move from their ground state, the minimum potential energy level, to an energized or excited state, a higher potential energy level (figure 1.20). The mechanism of this energization is uncertain but there are two postulated mechanisms: (1) resonance transfer and (2) exciton transfer. Electrons in the energized state are able to transfer their newly gained energy to another molecule within the pigment–protein complex and so on. Eventually the energy reaches the reaction centre chlorophyll, which in its excited state is able to lose an electron to a specific acceptor. In chloroplasts, P700 can pass an electron to ferredoxin and P680 can pass an electron to Q (section 2.8). This process leaves the reaction centre molecule without one electron, and unable to accept more energy. To allow more energy to be passed from the pigments the 'electron hole' is filled by an electron from a donor associated with the reaction centre concerned: plastocyanin for P700 and

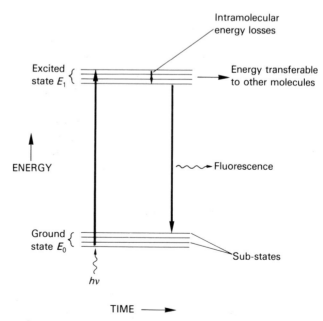

Figure 1.20   Movement of electrons between energy levels induced by light ($h\nu$)

water for P680. The energy absorbed by the chromophore is not always passed onto another pigment; sometimes a photon is released as the electron goes back to its ground state. This event can be observed as fluorescence and the photons emitted are always of longer wavelength (lower energy) than those absorbed, due to a loss of vibrational energy (as heat) during the energized state. Fluorescence is an essential overflow mechanism, occurring when light capture exceeds energy transference.

If the photosynthetic activity of chloroplasts at different wavelengths is measured, an 'action spectrum' can be constructed (figure 1.21), from which it

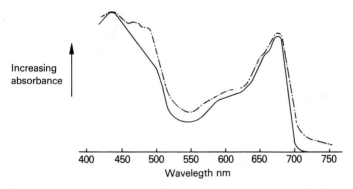

Figure 1.21   Action spectrum (—) and absorption spectrum (–·–··–) of a green alga. (Reproduced from *The Journal of General Physiology*, 1950, **33**, 404 by copyright permission of The Rockefeller University Press)

can be seen that the wavelengths most active in photosynthesis are almost exactly the same as those in the absorption spectrum of the pigments present. However, there is a major difference between the absorption spectrum and the action spectrum at wavelengths above 690 nm (far-red), where, although energy is captured, photosynthetic activity is lower than might be expected. This phenomenon is referred to as the 'red drop' effect. Emerson found that photosynthetic activity stimulated by these wavelengths (above 690 nm) could be greatly enhanced by using shorter wavelengths of light (673 nm) as well as the far-red. This was very important evidence for establishing the presence of two photosystems linked together in chloroplasts (section 1.5).

1.5 Functional organization of photosynthesis

The overall equation for photosynthesis in chloroplasts is

$$6CO_2 + 6H_2O \xrightarrow{h\nu^{\Delta}} C_6H_{12}O_6 + 6O_2$$

i.e., carbon dioxide and water are converted into glucose and oxygen is released. This is the sum of two separate sets of reactions which, for convenience, are called the 'light and dark' reactions. The light reactions are those which convert light energy to chemical energy and reducing power, viz., ATP and NAD(P)H, and release oxygen in the process, at least in chloroplasts. These take place in the thylakoid membranes of chloroplasts, and similar reactions occur in chromatophore membranes of photosynthetic bacteria. Light reactions are discussed in detail in chapter 2.

The dark reactions are considered to be those enzymes catalysed reactions which then use the ATP and NAD(P)H to fix CO_2 and form organic compounds. These occur in the chloroplast stroma and the photosynthetic bacterial cell cytoplasm. Dark reactions are discussed in detail in chapter 3. Light and dark reactions are intimately related: as well as the light reactions supplying energy for the dark reactions, light plays an important role in the control of the dark reactions (section 3.10).

The detailed history of the discovery of photosynthesis can be found in a large number of general textbooks. Here, only a few historical landmarks will be mentioned. In the 18th century Priestley showed that the presence of green plants in a sealed bell jar enabled mice to survive and a candle to continue to burn in that enclosed space. He identified the vital factor responsible for this survival as oxygen. That chloroplasts generated oxygen was concluded from the observation that oxygen-seeking bacteria migrate towards chloroplasts in the alga *Spirogyra*. It was also shown that light was necessary for the production of oxygen and for the fixation of CO_2 into organic matter. By isotopically labelling the oxygen of water it was demonstrated that the oxygen released by chloroplasts was derived from water and not CO_2. This and parallel studies on bacterial photosynthesis led Van Niel in the 1930s to put forward a general equation for photosynthesis, i.e.

$$CO_2 + 2H_2A \rightarrow (CH_2O) + H_2O + 2A$$
$$\text{carbohydrate}$$

So, in green plants, where water is a substrate and oxygen a product, this becomes:

$$CO_2 + 2H_2O \rightarrow (CH_2O) + H_2O + O_2$$

and in certain bacteria, where the substrate is H_2S:

$$CO_2 + 2H_2S \rightarrow (CH_2O) + H_2O + 2S$$

The idea that there were two sets of reactions in photosynthesis (light and dark reactions) came from the studies of action spectra (figure 1.21) and from the work of Blackman in the 1900s. He observed that at low light intensities, temperature had no effect on the overall rate of photosynthesis. At high light intensities, however, temperature did affect the rate. This implied that at low light intensities reactions involving light were rate-limiting, whereas at high light intensities some temperature-affected component was rate-limiting. The rates of reactions catalysed by enzymes are affected by temperature. Therefore, the conclusion was that there were light reactions and dark reactions, the latter catalyzed by enzymes.

In the 1930s Hill pioneered work on the compartmentation of photosynthetic reactions, which led to the now generally accepted view that the light reactions in the chloroplast occur in the thylakoid membranes while the dark reactions occur in the stroma. This work also established what is referred to as the Hill reaction:

$$H_2O + A \xrightarrow{\text{light}} AH_2 + \tfrac{1}{2}O_2$$

which represents a general equation for the light reactions. Hill used fragmented chloroplasts (grana preparations) which, in the presence of an artificial electron acceptor, e.g., ferricyanide, could evolve oxygen and reduce the electron acceptor. Ferricyanide can be reduced by an electron to give ferrocyanide.

$$Fe(CN)_6^{3-} + e^- \rightarrow Fe(CN)_6^{4-}$$

A in the Hill reaction is a hydrogen (and electron) acceptor. The reduction of A only occurs in the light and no CO_2 is fixed in these preparations. It was later discovered that $NADP^+$, which occurs naturally in chloroplasts, could replace the artificial electron acceptor.

$$NADP^+ + H_2O \xrightarrow{\text{light}} NADPH + H^+ + \tfrac{1}{2}O_2$$

The light reactions involve protein–pigment complexes in the thylakoid membrane or cell membrane of bacteria, which harvest the light, as briefly described earlier (section 1.4). There are a series of components in the membrane (some may not be in a complex) which are able to pass electrons from one component to the next by becoming alternately reduced then oxidized. This light-stimulated electron transfer is accompanied by the phosphorylation of ADP by inorganic phosphate (Pi) to give ATP. This is catalysed by the multi-subunit enzyme attached to the membrane (figure 1.13 and section 2.10), referred to as ATP synthetase (CF_1-ATPase).

The dark reactions are a cycle of enzyme-catalysed reactions which use ATP and NADPH to incorporate CO_2 into organic compounds and to produce the carbohydrate glucose. It was Calvin, Bassham and Benson in the 1950s who, by supplying chloroplasts with radiolabelled CO_2 and stopping the reaction after various time intervals, first identified (using 2-D chromatography and auto-radiography) the intermediates of the cycle and showed in what order they were produced.

A fundamental question about photosynthetic systems, which also applies to mitochondria and bacteria, was how could the electrical energy from the transfer of electrons, whether stimulated by light or by oxidative breakdown in the TCA cycle, be used for or coupled to the formation of the chemical energy 'store' ATP. A mechanism now generally accepted as the explanation, and called the chemiosmotic coupling theory, was first proposed in 1961 by Mitchell who, in 1979, was awarded the Nobel prize for his contribution to this field of research. The theory proposes that the electron transfer reactions bring about the formation of a hydrogen ion (H^+) gradient or membrane potential across the membrane and this is able to drive ATP synthesis (an endergonic reaction). A great deal of research into the coupling mechanism has been done using photosynthetic systems. A very elegant experiment carried out by Jagendorf (see figure 2.22) clearly demonstrates how a pH difference, i.e., H^+ gradient, could drive ATP synthesis. This is discussed in more detail in chapter 2.

1.5.1 Electron transfer

The light reactions involve a series of components, many of which are proteins containing chromophores, e.g., cytochromes contain haem, which are able to be reduced or oxidized as electrons are donated to them or passed on by them, respectively (reduction is the addition of electrons or hydrogens). For a cytochrome, where there is an iron atom in the centre of the haem group, the cytochrome is said to be oxidized or reduced according to the oxidation state of the iron atom. Two oxidation states are possible, ferrous (Fe^{2+}) and ferric (Fe^{3+}), which can be interconverted by the addition or subtraction of an electron:

$$Fe^{3+} + e^- \underset{\text{oxidation}}{\overset{\text{reduction}}{\rightleftharpoons}} Fe^{2+}$$

$$\underset{\substack{\text{oxidized} \\ \text{form}}}{} \qquad \underset{\substack{\text{reduced} \\ \text{form}}}{}$$

For another component of the light reactions, plastoquinone (PQ), the reduced and oxidized forms are interconverted as follows:

$$PQ + 2H^+ + 2e^- \underset{\text{oxidation}}{\overset{\text{reduction}}{\rightleftharpoons}} PQH_2$$

$$\underset{\substack{\text{oxidized} \\ \text{form}}}{} \qquad \underset{\substack{\text{reduced} \\ \text{form}}}{}$$

An oxidation reaction must always be accompanied by a reduction reaction. The following equation shows how these two adjacent components of the chloroplast electron transport chain can react, one (PQ) becoming oxidized and the other (cytochrome b) becoming reduced.

$$PQH_2 + 2\,cyt\ b_{559}(Fe^{3+}) \rightarrow PQ + 2H^+ + 2\,cyt\ b_{559}(Fe^{2+})$$

$$\underset{\text{reduced}}{} \quad \underset{\text{oxidized}}{} \qquad \underset{\text{oxidized}}{} \qquad \underset{\text{reduced}}{}$$

PQ/PQH_2 and cyt b_{red}/cyt b_{ox} are called redox couples and the reduction–oxidation potential (or redox potential) is a measure of the reducing or oxidizing power of a redox couple. A redox couple has a tendency to oxidize (accept electrons from) others of more negative potential and to reduce (donate electrons to) those of more positive potential. The standard redox potential (E^0) is a measure of the ability of a redox couple to donate electrons or to gain electrons under standard conditions, which are defined as pH 0 and unit activity (molar) of all components. The standard hydrogen electrode (H_2 gas at 1 atm pressure in equilibrium with H^+ ions in a solution of unit activity) has $E_0 = 0$. In fact this electrode potential is usually measured at pH 7 when $E_0 = -0.421$ V.

$$H^+ + e^- \rightleftharpoons \tfrac{1}{2} H_2$$

In practice, redox potentials in systems like chloroplasts, photosynthetic bacteria and mitochondria cannot be measured under standard conditions and so the mindpoint potential (E'_0) is measured using more physiological conditions like temperatures of 25°C and 30°C and at pH 7, and when the concentrations of the oxidant and reductant are equal. The midpoint potential is measured by placing the sample under investigation in a very special cuvette, in a dual wavelength spectrophotometer under anaerobic conditions. Redox dyes are present to allow the potential to change gradually as a reductant or oxidant is added in small amounts. The midpoint potential is determined using the Nernst equation:

$$E_{\text{obs}} = E'_0 + \frac{RT}{nF} \; 2.303 \log_{10} \frac{[\text{electron acceptor}]}{[\text{electron donor}]}$$

where R is the gas constant (8.31 J $\deg^{-1}$ mol^{-1}), T is the temperature in degrees Kelvin (25°C = 298°K) the absolute temperature, n represents the number of electrons transferred per mole (for PQ $n = 2$, for a cytochrome $n = 1$) and F is the Faraday (96400 J V^{-1}). The square brackets mean 'the molar concentration of'. When the component whose midpoint potential is to be measured is 50 % reduced, i.e., the amount of electron donor equals the amount of electron acceptor, then the observed potential (E_{obs}) becomes equal to the midpoint potential (E'_0). Midpoint potentials have been used to determine the approximate order of components in electron transfer chains.

1.5.2 Pigment–protein complexes

The thylakoid membrane probably contains about 30–50 different proteins (see chapter 4) and the membrane has been fractionated following solubilization with detergents, primarily SDS (sodium dodecyl sulphate) and then electrophoresis, into several complexes containing pigments, e.g., chlorophyll, and proteins. These are referred to as chlorophyll–protein (CP) complexes. Clear details about these complexes are gradually emerging as this is currently a very active research field. There appears to be one complex associated with photosystem 1 (containing P700 and chlorophyll a) which is able to reduce $NADP^+$ in the presence of artificial electron donors. There is also a complex associated with photosystem 2 (containing P680, chlorophyll a and some chlorophyll b) which is able to carry out the Hill reaction with ferricyanide as electron acceptor. Another complex, referred to as the light-harvesting complex (LHC), contains about half of the chlorophyll a in the thylakoid membrane and nearly all the chlorophyll b. This complex may represent two (or more) complexes, one associated with photosystem 1 and one with photosystem 2. There is apparently no light-harvesting complex in cyanobacteria, which do not have any chlorophyll *b*.

Another complex associated with thylakoid membranes is the multi-subunit complex which makes ATP. It includes a membrane section (CF_0) and the ATPase or coupling factor (CF_1) but no pigments.

Freeze-fracture studies of thylakoid membrane preparations have shown different sized proteins associated with the different fracture faces of the membranes and work is proceeding to try to find how these relate to the complexes and to the amount of thylakoid stacking to form grana. Figure 1.22

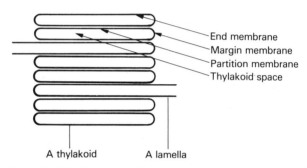

Figure 1.22   Diagrammatic representation of a granum stack

shows that the grana membranes have different immediate environments: the end membranes and margins are in contact with the stroma, but those in the partition are not. Techniques devised to separate stroma-facing membranes from partition membranes have allowed researchers to determine the position of various proteins and complexes within these membranes. It has been postulated that the photosystem 1 complex and the ATP synthetase complex are situated mainly in the stroma, facing membranes, while the photosystem 2 complex is associated mainly with the partition membranes. Cytochrome *b* and *f*, which connect the two photosystems, are distributed fairly evenly throughout the membranes. It has been suggested that plastocyanin and plastoquinone act as mobile electron carriers and, with cytochromes *b* and *f*, connect the two photosystems, allowing electrons to pass from photosystem 2 to photosystem 1. It is interesting to note that bundle sheath chloroplasts (figure 1.8), which contain very few grana and therefore have a low level of partition membranes, also have low levels of photosystem 2.

The amount of stroma-facing membrane also depends on the type of plant from which the chloroplast comes. Shade plants have very large grana stacks and have lower ratios of stroma-exposed membranes to partition membranes than Sun plants. Shade plants can achieve only low rates of photosynthesis per unit of chlorophyll, which become saturated at low light intensities. Sun plants, however, can achieve higher rates of photosynthesis per unit of chlorophyll. These observations would be consistent with there being lower levels of photosystem 1 and ATP synthetase in Shade plants, due to their lower levels of stroma-facing membrane, than in Sun plants.

1.6 Preparation of photosynthetic systems

Photosynthetic organisms are normally grown in axenic (pure) cell culture if possible; plants can be grown in growth cabinets where conditions can be controlled. Cell culture allows the cells to be kept in a medium where the concentrations of the carbon source, inorganic salts, trace elements, vitamins and conditions of temperature and light can be fully controlled. Growing samples in the dark and then exposing them to light is especially useful for observing the effects of light on the control of metabolism. Large amounts of material, e.g., 100 litres, can be grown under cell culture. The cells can be collected using a continuous flow centrifuge: the cell suspension is continuously fed into the rotor during centrifugation causing the cells to collect in a special chamber. This method keeps the cells relatively free of contamination.

1.6.1 Preparation of higher plant chloroplasts

Spinach (*Spinacia oleracea*) is the (C_3) plant very often used as a source of chloroplasts although peas and lettuce are frequently used as well. The first stage in the preparation may involve the formation of protoplasts, i.e., the removal of the very strong cell wall, leaving the cell membrane intact. This is invariably done before isolating chloroplasts from mesophyll and bundle sheath cells from C_4 plants. Cell walls are often removed first as the force needed to rupture them could also disrupt the fairly fragile cell contents, e.g., chloroplasts and other organelles. Protoplasts are formed by enzymic digestion of the cell wall using enzymes like cellulases, pectinases and hemicellulases, which can be isolated from microbes such as *Trichoderma viride* or *Aspergillus niger*. After this digestion of the cell wall the cell membrane requires less force to break it.

The chloroplast preparation is carried out very quickly. Figure 1.23 shows a schematic representation of the method. Cooled leaves are ground in a blender in sorbitol medium with a concentration of about 0.33 M, at pH 7.5 for 3–5 seconds. The mixture is then passed through two layers of muslin and then filtered through eight layers of muslin and some cotton wool. The filtrate is then

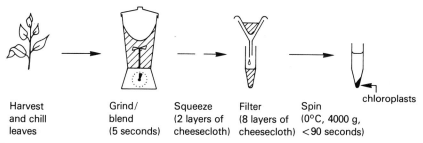

| Harvest and chill leaves | Grind/ blend (5 seconds) | Squeeze (2 layers of cheesecloth) | Filter (8 layers of cheesecloth) | Spin (0°C, 4000 g, <90 seconds) | chloroplasts |

Figure 1.23   Flow diagram showing the preparation of higher plant chloroplasts: all procedures carred out at 4°C.

placed in a centrifuge and taken from 0 to 6000 rpm and back in 90 seconds. The pellet contains the chloroplasts, which may be resuspended using a little cotton wool on a rod in medium of pH 7.5. These chloroplasts are called class I chloroplasts, and are capable of high rates of CO_2 fixation.

1.6.2 Preparation of chloroplast envelope

Chloroplast envelope preparations are made by rupturing isolated whole chloroplasts, by exposing them to hypotonic media. The envelope fraction can then be separated out by differential centrifugation or centrifugation on a density gradient. The envelope preparation contains vesicles which have only one membrane, some outer membrane and some inner membrane vesicles.

Thylakoid membranes can be isolated in a similar way; the chloroplast is disrupted with hypotonic media or detergents such as digitonin, and then the membranes are separated out by centrifugation.

1.6.3 Preparation of chromatophores

Chromatophore preparations from photosynthetic bacteria are the broken off invaginations of the inner membrane which have resealed. The bacterial cells, collected as described earlier, are resuspended in a small amount of buffer at pH 7.5. The cells are then disrupted either by passing the suspension through a precooled French pressure cell or by freezing and thawing the suspension several times. Differential centrifugation at 4°C and 16 000 g for 20 minutes then gives a pellet containing cell debris, which is discarded. The supernatant which contains the chromatophores is then centrifuged again at 177 000 g for 60 minutes. The pellet from this centrifugation contains the chromatophores.

1.7 Photosynthesis in the whole plant

The investigation of the detailed history of photosynthetic systems has necessitated the use of isolated chloroplasts, subchloroplast preparations, chromatophores, etc. This has tended to cloud the importance of looking at the process of photosynthesis with respect to the plant as an individual or as a member of a plant community. This is clearly of great importance when one is attempting to maximize photosynthetic production as in a crop plant, for example. Efficient photosynthesis requires delivery to the photosynthetic re-action centres of sufficient energy (ultimately derived from light energy) to provide the driving force for the light reactions of photosynthesis, and the delivery to the chloroplast stroma of adequate amounts of basic raw materials of photosynthesis – carbon dioxide and water.

The interaction of light with plants is complex. The dependence of whole plant photosynthesis on light intensity varies with temperature and the avail-ability of carbon dioxide. In most plants light levels govern photosynthetic rate only up to a particular intensity of light. Above this intensity the overall rate is governed by the light-independent phase of photosynthesis (the dark reactions), and these are limited either by temperature or CO_2 availability. In the case of many plants, the point at which light intensity ceases to limit photosynthetic productivity can be below the level of normal sunlight. In other plants, normal sunlight levels still limit the photosynthetic rate as a whole. It is consequently of great importance in these latter cases to maximize the light impinging on the leaf surface. This is achieved in part by the positively phototropic properties of the plant stem, pushing the leaves as high as possible and clear of those of competitors, and in part by the fact that most leaves exhibit diaphototrophism, orienting themselves to receive the maximum light flux. Such movements probably also benefit plants in a situation where temperature is limiting photosynthetic rate.

In natural conditions carbon dioxide availability is often the rate-limiting factor in photosynthetic productivity. The external surfaces of the leaves of higher plants are usually covered with a coating of waxy material which is relatively impermeable to the photosynthetic gases carbon dioxide and oxygen, as well as water. Delivery of atmospheric CO_2 to the stroma of the chloroplast from outside the plant depends on the presence of stomata in the leaf surfaces. Stomata are pores which are bordered and controlled by a pair of modified epidermal cells known as guard cells. The pore is closed when the guard cells are flaccid and open when they are turgid. When photosynthetic activity is high, the osmotic potential of the photosynthate produced in the guard cells causes water uptake, turgidity and stomatal opening. This facilitates entry of CO_2 into the air spaces of the leaf surrounding the photosynthetic mesophyll cells to replenish that previously used in photosynthesis.

The opening of the stomatal pores has the added effect of permitting water loss from the leaf to the atmosphere. This is replenished by water from the soil entering the roots and moving up the vascular system of the plant (transpiration stream – xylem). As a general rule, the CO_2 uptake and water loss are maintained in balance by these mechanisms. However, many plants which exist in extremely hot and dry conditions exhibit further structural and biochemical adaptations to maximize CO_2 uptake and minimize dehydration from transpiration. The first of these is termed Crassulacean acid metabolism (CAM) (section 3.4) and plants exhibiting this feature are usually very succulent with either thick fleshy leaves or with leaves modified to spines and a fleshy swollen photosynthetic stem. To conserve water these plants have stomata which close in the day and open at night. Consequently during the daytime, when photosynthesis is possible, the plant does not have access to significant quantities of atmospheric CO_2. At night CO_2 enters the leaves through the stomata and reacts with phosphoenol pyruvate (PEP) in the cytosol of the mesophyll cells to form oxaloacetate and ultimately malate, which is transported to, and stored in, the vacuoles. It is thought that the fleshy nature of these plants results from increased vacuolation to maximize CO_2 storage. During the day malate is returned from the vacuole to the cytosol and decarboxylated to produce CO_2 necessary for photosynthesis.

Another means for conserving CO_2 occurs in plants having what is referred to as a C_4 mechanism of photosynthesis (section 3.3). The C_4 biochemistry is usually associated with structural modifications to the leaf known as Kranz anatomy. 'Kranz' is a German word meaning 'wreath' and describes the ring of cells with prominent development of modified chloroplasts and forming a sheath around the vascular bundles of the leaves of these plants. As in the CAM plants CO_2 is trapped by reaction with PEP to form oxaloacetate, which is reduced to malate in the cytosol of the mesophyll cells. In this case, however, malate is transported to the bundle sheath cells where CO_2 is released and used for photosynthesis. The mechanism brings about an increased concentration of CO_2 in the bundle sheath cells. The CO_2 entrapment in the mesophyll cell cytosol has further significance. Loss of water under hot dry conditions results in the flaccidity of the guard cells and closure of the stomatal pores. The low uptake of CO_2 under these conditions demands its maximal conservation. The cytosolic CO_2 trapping system is thought to mop up and reduce loss from the plant of CO_2 released in photorespiration (section 3.6). It also brings about a lowering of the partial pressure of CO_2 in the air spaces of the leaves, increasing the CO_2 gradient between the inside and outside of the plant.

Plants grown in bright sunshine (Sun plants) are genetically adapted to the light intensity difference. Shade plants have larger chloroplasts, less soluble protein, but more chlorophyll per cell. Their chloroplasts have large grana stacks and contain higher proportions of chlorophyll b to chlorophyll a than Sun plants. There may be correlation between number of grana stacks and the ratio of the chlorophylls as it is thought that grana thylakoids have higher ratios of chlorophyll b to a than those in the stroma.

Overall, photosynthetic rates of chloroplasts from Shade plants are lower than those of Sun plants, probably due to lower levels of electron transport intermediates such as cytochromes, and of ribulose bisphosphate carboxylase (RBPC), which is the enzyme that fixes CO_2 in the Calvin cycle.

Suggested further reading

Books

Carr, N. G. and Whitton, B. A. (Eds.) (1983). *The Biology of Cyanobacteria*, 2nd Edn. Oxford University Press, London
Dodge, J. (1973). *The Fine Structure of Algal Cells*, Academic Press, London
Gregory, R. P. F. (1977). *Biochemistry of Photosynthesis*, Wiley, Chichester
Hall, D. O. and Rao, K. K. (1972). *Photosynthesis*, Studies in Biology, no. 37, Edward Arnold, London
Harrison, R. and Lunt, G. G. (1980). *Biological Membranes*, 2nd ed., Tertiary Level Biology, Blackie, London
Reid, R. A. and Leech, R. M. (1980). *Biochemistry and Structure of Cell Organelles*, Blackie, London
Robards, A. W. (1970). *Electron Microscopy and Plant Ultrastructure*, McGraw-Hill, Maidenhead

Tribe, M. A. and Whittaker, P. A. (1981). *Chloroplasts and Mitochondria*, 2nd ed., Studies in Biology, no. 31, Edward Arnold, London

Whatley, J. M. and Whatley, F. R. (1980). *Light and Plant Life*, Studies in Biology, no. 124, Edward Arnold, London

Articles

Anderson, J. M. (1975). The molecular organisation of chloroplast thylakoids, *Biochem. Biophys. Acta*, **416**, 191

Anderson, J. M. and Andersson, B. (1982). The architecture of photosynthetic membranes: lateral and transverse organisation. *Trends Biochem. Sci.*, **7**, 288

Barber, J. (1982). Influences of surface charge on thylakoid structure, *Ann. Rev. Plant Physiol.*, **33**, 261

Boardman, N. K. (1977). Comparative photosynthesis of Sun and Shade plants. *Ann. Rev. Plant. Physiol.*, **28**, 355

DePierre, J. W. and Ernster, L. (1977). Enzyme topology of intracellular membranes, *Ann. Rev. Plant Physiol.*, **46**, 201

Heber, U. and Heldt, H. W. (1981). The chloroplast envelope. Structure, function and role in leaf metabolism. *Ann. Rev. Plant Physiol.*, **32**, 139

Hill, R. (1965). The biochemist's green mansions: the photosynthetic electron transport chain in plants. *Essays in Biochem.* (Eds. P. N. Campbell & G. D. Greville), Vol. 1, 121

Singer, S. J. and Nicolson, G. L. (1972). The fluid mosaic model of the structure of cell membranes. *Science*, **175**, 720

Zaborsky, O. R. *et al.* (1982). *Handbook of Biosolar Resources: Fundamental Principles*, Vol. 1, CRC Press, New York, p. 1

2
Photosynthetic Phosphorylation

2.1 Introduction

In chapter 1 we have seen that a variety of prokaryotes and eukaryotes may derive energy from sunlight. This energy is used to drive photosynthetic electron transport which produces ATP and NADH or NADPH to be used for carbon and nitrogen fixation (see sections 3.2 and 3.8). This chapter describes the first half of this story – from light absorption to the production of ATP and reducing power – the 'light reactions'. The chapter starts with a general consideration of light absorption and the primary photosynthetic reactions, followed by a description of electron transport and ATP synthesis. Later examples are looked at from both prokaryotic and eukaryotic organisms of specific energy conserving systems, based on the general principles learnt in the first part of the chapter. Finally, consideration is given to some more controversial topics such as efficiency and regulation of the light reactions of photosynthesis.

2.2 Pigments and the absorption of light

In order to be able to utilize the sun's light as a source of energy, an organism must be able to absorb a proportion of that light. This absorption mechanism involves different pigments, which are responsible for the variety of colours of photosynthetic organisms which, in turn, lend colour to the Earth's landscape. The pigments found in photosynthetic organisms fall into three main groups: chlorophylls, carotenoids and phycobilins.

2.2.1 Chlorophylls

Chlorophylls are found in all photosynthetic organisms. The active, common prosthetic group of the molecule is a magnesium-porphyrin ring which confers the greenish colour characteristic of all chlorophylls. Chlorophyll structure is summarized in figure 2.1. In all higher plants and green algae chlorophyll a and chlorophyll b are both found, but in some eukaryotic algae chlorophyll b may be replaced with chlorophylls c and d. The chemical differences are manifested in the differing absorption peaks, shown in figure 2.2 and table 2.1. Photo-

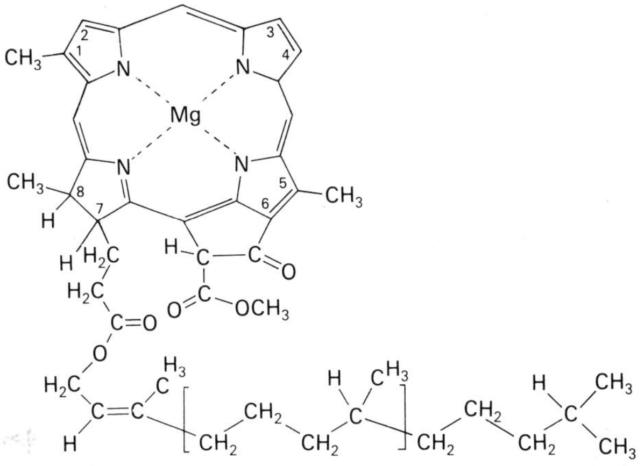

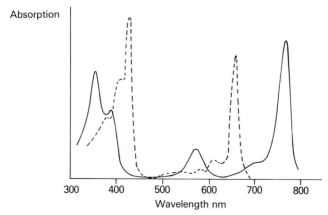

Position	2	3	4
Chl *a*	—CH=CH$_2$	—CH$_3$	—C$_2$H$_5$
Chl *b*	—CH=CH$_2$	—CHO	—C$_2$H$_5$
Bchl *a*	—COCH$_3$	—CH$_3$	—C$_2$H$_5$
Bchl *b*	—COCH$_3$	—CH$_3$	=CH—CH$_3$

Figure 2.1   The structure of chlorophylls. Chl: chlorophyll, Bchl: bacteriochlorophyll

synthetic bacteria contain bacteriochlorophylls which absorb light at much longer wavelengths than chlorophyll *a*. All photosynthetic bacteria contain bacteriochlorophyll *a* and, in addition, the purple sulphur bacteria (Chromatiaceae) and the non-sulphur bacteria (Rhodospirillaceae) may contain bacteriochlorophyll *b*. The green sulphur bacteria (Chlorobiaceae) contain bacteriochlorophylls *c*, *d* and *e* in addition to bacteriochlorophyll *a*.

Absorption

300   400   500   600   700   800
Wavelength nm

Figure 2.2   Absorption spectra of chlorophyll *a* and bacteriochlorophyll in ethanolic solution. —— bacteriochlorophyll, ------ chlorophyll *a*

Table 2.1 Absorption and fluorescence maxima of chlorophylls

Chlorophyll	Wavelength absorption (nm)	Maxima in ether (nm) fluorescence
a	428.5, 660	668, 723
b	452.5, 642.0	649, 708
c	444.4, 628.2	
d	447, 688	693, 705
Bchl	357.6, 770	695, 805
Chlorobium 660	430, 660	667

The amount of energy from sunlight made available to the organism to do work is determined by the absorption peaks of the reaction centre chlorophylls (see section 1.4).

2.2.2 Pheophytins

These chlorophyll-type pigments that lack the central Mg in the pyrrole ring are also found in photosynthetic organisms. They are known as pheophytin a, b, bacteriopheophytin, etc. Originally they were thought to be breakdown products of the chlorophylls, but recently they have been shown to take an important part in photosynthesis (see section 2.8.1).

2.2.3 Carotenoids

A large number of different carotenoids are present in photosynthetic organisms, some being typical of algae and higher plant chloroplasts and others of photosynthetic bacteria (figure 2.3). Carotenoids are usually, red, brown,

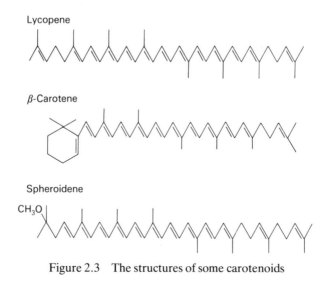

Lycopene

β-Carotene

Spheroidene

CH₃O

Figure 2.3   The structures of some carotenoids

orange or yellow, and the brown or pink colour of certain photosynthetic bacteria is conferred by the presence of carotenoids. The absorption spectrum of β-carotene is shown in figure 2.4. Chemically, carotenoids are considered to be a class of hydrocarbons (carotenes) and their oxygenated derivatives (xanthophylls) consisting of eight isoprenoid units joined in such a manner that the arrangement of isoprenoid units is reversed at the centre of the molecule so that the two central methyl groups are in a 1,6 positional relationship. All carotenoids may be formally derived from the basic structure of lycopene (figure 2.3a) by hydrogenation, dehydrogenation, cyclization or oxidation or any combination of these processes.

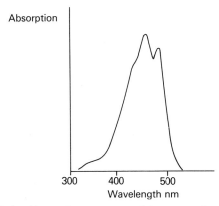

Figure 2.4   Absorption spectrum of β-carotene in petroleum

2.2.4 Phycobilins

Phycobilins are far less common in the photosynthetic world than chlorophylls or carotenoids. Phycocyanins are blue and are found in cyanobacteria (blue-green algae), while red phycoerythrins may be found in some cyanobacteria and red eukaryotic algae. The structures of some phycobilins are shown in figure 2.5 and absorption spectra in figure 2.6.

2.3 The functions of pigments – light-harvesting complexes

As we said in the last section, the primary function of pigments is to absorb or 'harvest' the sunlight. Specialized pigment–protein complexes (section 1.5.2), associated with the photosynthetic membrane, light-harvesting complexes, perform this function rather than just the isolated pigments. The arrangement of pigments within these complexes and the arrangement of one complex with respect to the other is very important and enables energy to be transferred between individual pigments with a very high efficiency. The mechanism of energy transfer between pigments within one light-harvesting complex is very much like that found in crystals, reflecting the peculiarly ordered arrangement

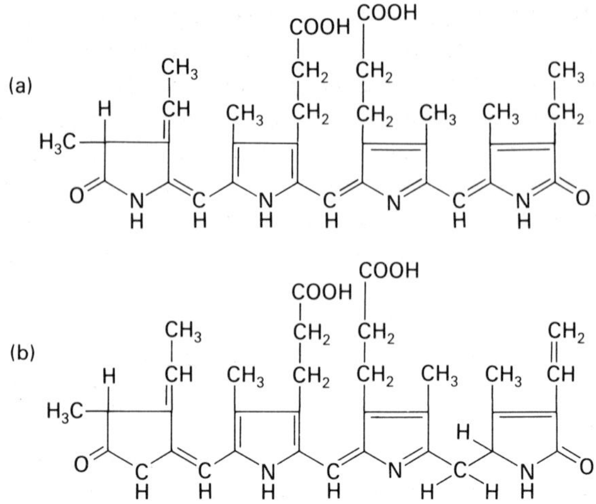

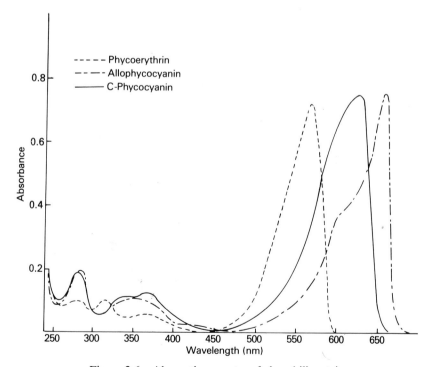

Figure 2.5   The structure of phycobiliproteins. All phycobiliproteins contain one of the chromophores (a) or (b) covalently bound to protein. (a) Phycocyanobilin, chromophore of the phycocyanins (blue); (b) phycoerythrobilin, chromophore of the phycoerythrins (red)

Figure 2.6   Absorption spectra of phycobiliproteins

of chlorophylls. In higher plant chloroplasts, light harvesting is performed by three complexes; a chlorophyll *a*-protein, chlorophyll *b*-protein and a chlorophyll *a* + *b*-protein which are all integral parts of the membrane. In cyanobacteria the main light-harvesting complex is a specialized structure called a phycobilisome, attached to the membrane, containing phycobilins. A similar specialized structure, containing chlorophylls *c*, *d* and *e* is found in the green sulphur bacteria, while in the purple bacterial light-harvesting complexes, either bacteriochlorophyll *a* or *b*-proteins, are integral parts of the membrane.

Thus the light-harvesting complexes of photosynthetic organisms contain a variety of different pigments, allowing them to use a large proportion of the visible spectrum. Figure 2.7 shows the absorption spectra of the organisms mentioned above.

Carotenoids are found in all light-harvesting complexes which also contain chlorophyll. Although they play a part in absorbing sunlight, they also perform a very important function in protecting chlorophylls from destruction by sunlight.

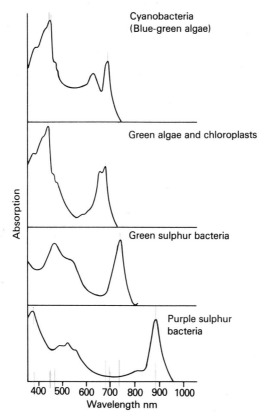

Figure 2.7   Absorption spectra of photosynthetic organisms

2.4 The functions of pigments – the reaction centre

All energy absorbed by pigments in light-harvesting complexes is transferred to a specialized complex of pigments and proteins known as a photosynthetic reaction centre (figure 2.8). The efficiency of transfer of energy is about 98 %, due to the special orientation of the light-harvesting complexes with respect to the reaction centre. The energy arriving at the reaction centre is used to oxidize a special chlorophyll or chlorophyll dimer. Chlorophyll is normally considered to be an oxidizing agent itself, with a midpoint redox potential (section 1.5.1) of

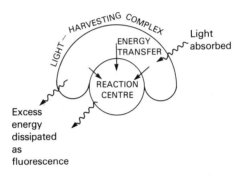

Figure 2.8   Energy transfer from the light-harvesting complex to the reaction centre

about 450 mV, so the loss of an electron requires a considerable amount of energy. The electron is transferred to a 'primary electron acceptor' with a much lower redox potential than chlorophyll, i.e., normally this component would be considered to be a reducing agent. The 'hole' or positive charge left in the special chlorophyll is filled by an electron from a 'primary electron donor'. These electron transfer reactions are extremely fast, occuring in picoseconds (10^{-12} s), and can still be measured down to 4°K. Oxidation of the primary electron acceptor via an electron transport chain drives ATP synthesis.

When the primary electron acceptor is already reduced, no electrons can be transferred from the specialized reaction centre chlorophyll to the acceptor. Under these conditions this chlorophyll loses its energy either as heat or fluorescence, i.e., the re-emission of light. Thus fluorescence of the reaction centre can be seen as an overflow mechanism for the energy absorbed by the light-harvesting pigments from sunlight. All photosynthetic organisms emit a low level of fluorescence under normal conditions because electron transfer away from the primary electron acceptor is slower than the excitation of the reaction centre chlorophyll. These reactions are summarized in figure 2.9.

There are three types of reaction centre in photosynthetic organisms: the bacterial reaction centre, photosystem 1 and photosystem 2. They will be discussed in more detail later in the chapter. The bacterial reaction centre has bacteriochlorophyll *a* as its specialized chlorophyll. Photosystems 1 and 2 are found in all chloroplasts and cyanobacteria, and the specialized chlorophyll is chlorophyll *a*.

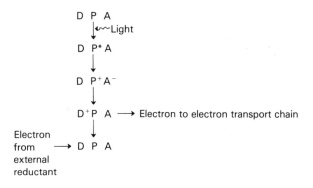

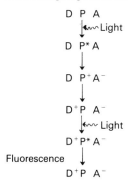

Conditions giving rise to fluorescence:

Energy of P* cannot be used to form P^+ because A is already reduced.

Figure 2.9 The photosynthetic reaction centre. P: reaction centre chlorophyll, A: primary electron receptor, D: primary electron donor

2.5 Oxygenic and anoxygenic photosynthesis

As we have already stated in chapter 1, the end products of photosynthetic electron transport are NADH or NADPH and ATP which are used for metabolism in the cell. ATP is synthesized as a consequence of the reoxidation of the primary electron acceptor. In order that NAD^+ or $NADP^+$ may be reduced, a source of reductant must be provided to the cell, to re-reduce the primary electron donor. For photosynthetic bacteria this reductant may be H_2, H_2S or a carboxylic acid such as succinate. For example, when H_2S is oxidized, sulphur is produced:

$$H_2S \rightarrow 2H^+ + 2e^- + S$$

In chloroplasts and cyanobacteria, water is used as the reductant:

$$2H_2O \rightarrow 4H^+ + 4e^- + O_2$$

Oxygen is given off, and therefore this is known as oxygenic photosynthesis.

2.6 Electron transfer carriers

ATP is synthesized in bacteria, mitochondria and chloroplasts by coupling the ATP synthetase enzyme to the oxidation of a pool of reductant. The latter is oxidized by a stepwise mechanism in an electron transport chain composed of carriers which can themselves be readily oxidized and reduced. The mediators found in all organisms fall into the following groups: quinones, flavoproteins, iron–sulphur proteins, cytochromes, plastocyanin.

2.6.1 Quinones

Ubiquinone is found in bacteria and mitochondria, while plastoquinone is present in chloroplasts, as large pools fixed within the membranes. Quinones

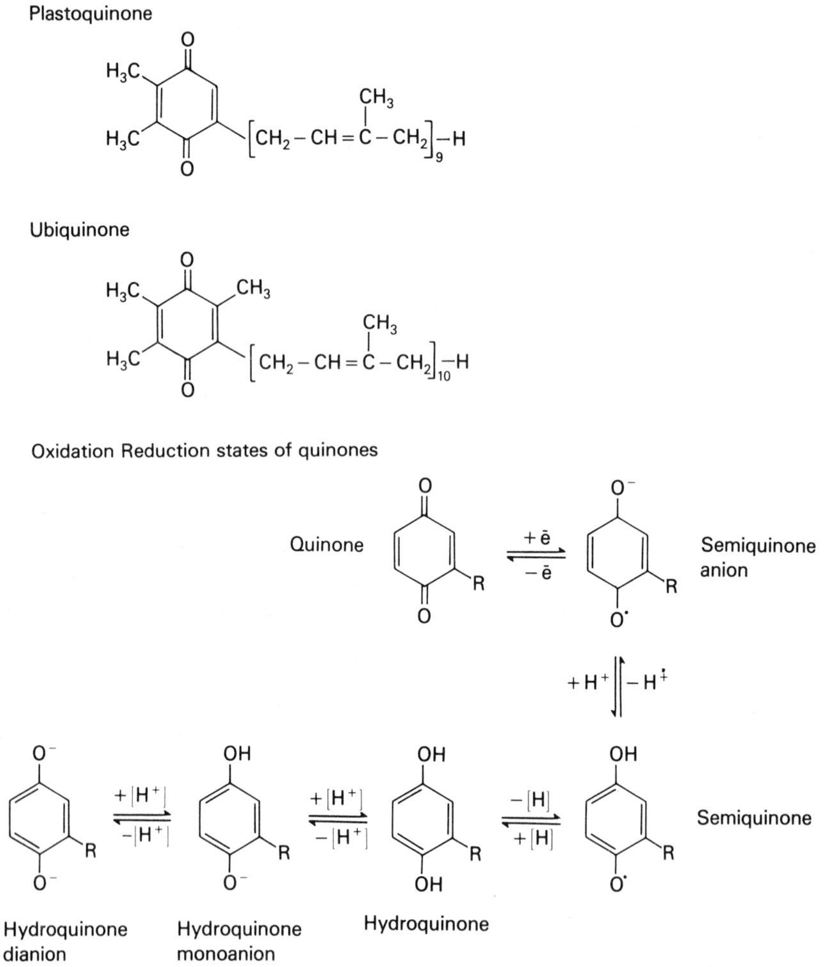

Figure 2.10   Quinones

are usually considered to be 'hydrogen carriers' because of the normal oxidation/reduction reaction (see figure 2.10):

$$Q + 2[H] \rightleftharpoons QH_2$$

However, other quinone species probably exist in electron transport reactions, such as the semiquinone anion. These are shown in figure 2.10. These reduced quinones show differing absorption spectra in the ultraviolet (figure 2.11).

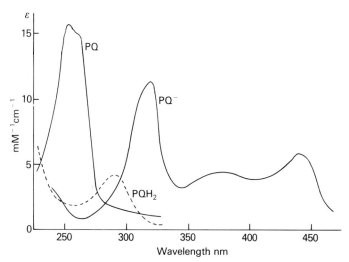

Figure 2.11   Absorption spectra of plastoquinone. (From *Encyclopaedia of Plant Physiology*, volume 5, chapter 13, Plastoquinone by J. Amesz, Springer, New York, 1977)

2.6.2 Flavoproteins

Flavoproteins are found in all electron transport chains and have the prosthetic group riboflavin, shown in figure 2.12. The possible oxidation/reduction reactions are very similar to those of quinones, but in electron transport chains flavoproteins act as hydrogen carriers undergoing the reactions:

$$Fp + 2[H] \rightleftharpoons FpH_2$$

The typical absorption spectra of flavoproteins is shown in figure 2.13.

Flavoproteins are aqueously soluble enzymes and therefore comparatively easy to extract and characterize. One particularly well characterized flavoprotein is the NADP-ferredoxin reductase of chloroplasts and algae, which we shall mention later.

2.6.3 Iron–sulphur proteins

These proteins are considered to be the most primitive of all electron transport mediators, being relatively low molecular weight proteins and ubiquitous in

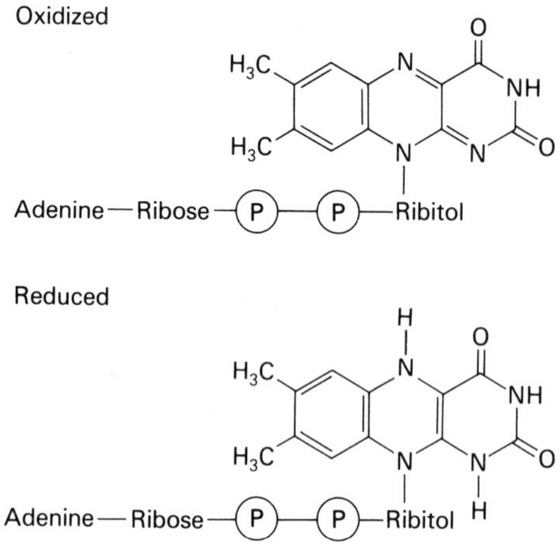

46

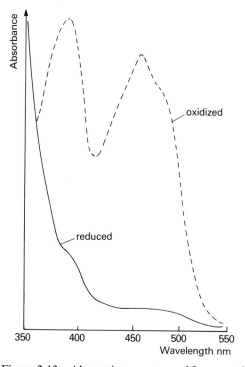

Oxidized

Reduced

Figure 2.12   Flavin adenine dinucleotide: the flavoprotein prosthetic group of NADP-
ferredoxin reductase

Figure 2.13   Absorption spectrum of flavoprotein

nature. A large number of these proteins have been investigated in depth and the primary amino acid sequence of many such iron–sulphur proteins has been determined. By comparing these sequences it is possible to propose an evolutionary relationship between organisms.

Iron–sulphur proteins may either be soluble in aqueous medium or membrane bound. Three classes of protein exist: (1) 8Fe-8S, (2) 4Fe-4S, (3) 2Fe-2S. The 8Fe-8S proteins are found in relatively primitive bacteria, the 4Fe-4S proteins are found more generally and the best known, 2Fe-2S protein, is the soluble ferredoxin of chloroplasts. The proposed models of a 2Fe-2S centre and a 4Fe-4S centre are shown in figure 2.14. The 8Fe-8S iron–sulphur proteins contain two 4Fe-4S centres. Iron–sulphur centres are one electron acceptors, the change in the redox state being distributed between the iron atoms (see figure 2.14). These electron carriers have been characterized using

(a)

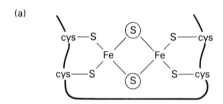

(b)

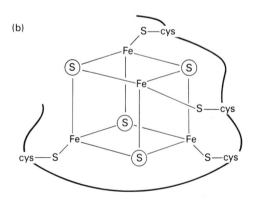

Figure 2.14 Iron–sulphur centres: (a) 2FE-2S iron–sulphur centre, (b) 4FE-4S iron–sulphur centre. The thick line is protein

several different techniques including absorption spectra (see figure 2.15), electron paramagnetic resonance, endor, nuclear magnetic resonance and Mössbauer spectroscopy. This concerted effort has been well reviewed in references quoted at the end of this chapter, and represents an example of the way in which many varied techniques can be used to solve biochemical problems.

48

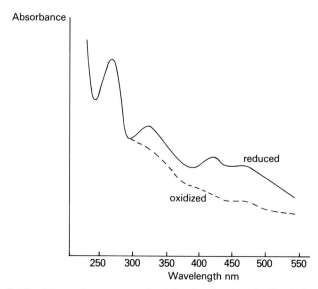

Figure 2.15   Absorption spectra of oxidized and reduced spinach ferredoxin

2.6.4 Cytochromes

Cytochromes also have prosthetic groups containing iron, but coordinated to a porphyrin ring as haem (figure 2.16). Cytochromes may be soluble or membrane bound, and may vary in molecular weight from 10 000 to 40 000. The visible absorption spectrum (see figure 2.17) is very characteristic and the absorption bands between 545 nm and 565 nm, the α peaks, have often been used as the main characteristics of certain cytochromes, e.g., cytochrome b_{559} of chloroplasts. Cytochromes are classified into classes a, b, c, d, e and o by the different chemical groups attached to the haem ring (figure 2.16). Only cytochromes in the b and c group are found in photosynthetic electron transport chains, some of which are listed in table 2.2.

2.6.5 Plastocyanin

Plastocyanin is a copper protein unique to chloroplasts. The copper is co-ordinated as shown in figure 2.18a to give a characteristic blue colour, the absorption spectrum being that shown in figure 2.18b. The redox couple is the gain or loss of one electron so:

$$Cu^+ \rightarrow Cu^{2+} + e^-.$$

2.7 The function of carriers – electron transport

As we have said previously, the function of photosynthetic electron transport is to produce NADH or NADPH and ATP which can be used for carbon

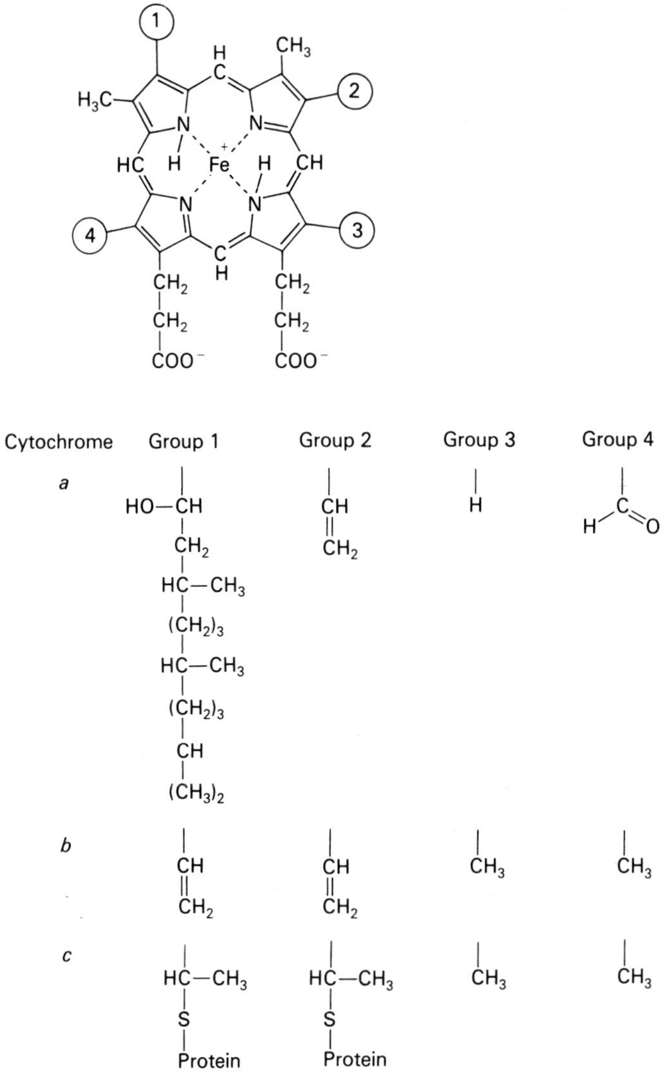

Figure 2.16   The structure of cytochromes

Cytochrome	Group 1	Group 2	Group 3	Group 4
a	HO—CH CH₂ HC—CH₃ (CH₂)₃ HC—CH₃ (CH₂)₃ CH (CH₃)₂	CH ‖ CH₂	H	H—C=O
b	CH ‖ CH₂	CH ‖ CH₂	CH₃	CH₃
c	HC—CH₃ S Protein	HC—CH₃ S Protein	CH₃	CH₃

or nitrogen fixation and for intermediary metabolism, which is discussed in chapter 3.

2.7.1   The mechanism of electron transport-linked phosphorylation

Electron transport carriers become reversibly oxidized and reduced in a stepwise mechanism in the order of their 'reducibility'. This 'reducibility' is measured by the mid-point redox potential of the carrier, a scale which has

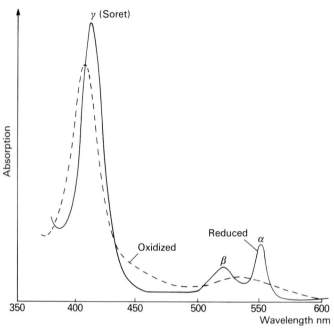

Figure 2.17 Absorption spectra of cytochromes. A typical spectrum: cytochrome *c* from beef heart mitochondria

Table 2.2 Photosynthetic cytochromes

Cytochrome	Source	MW/haem.	Maximal absorbance (nm) α β γ	Redox potential (v)
f	Parsley	61–68,000	554, 525, 423	0.365
f	Algae	16,500–23,000	552, 520, 415	0.340–0.390
			556, 521, 419	
b_6 (b_{563})	Spinach	40,000	563	−0.08
b_{559}	Spinach	46,000	559	various
				(see section 2.8.5)
c_2	*Rps. sphaeroides*	14,000	551, 521, 417	0.340

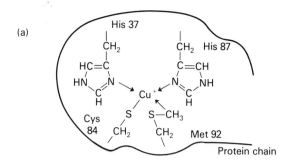

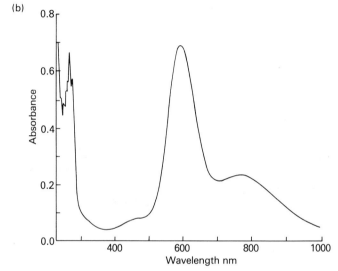

Figure 2.18 Plastocyanin: (a) the coordination of copper at the active centre; (b) absorption spectrum of spinach plastocyanin. (From J. Amesz, *Encyclopaedia of Plant Physiology*, Springer, New York, 1977, p. 248)

hydrogen at pH 0.0 as zero and -0.421 V at pH 7.0. Figure 2.19 shows a typical scale for some cytochromes, iron–sulphur centres and quinones. Thus in theory we could construct an electron transport chain as shown in figure 2.20a. In general, this order of carriers is what is found in nature, but some complications with respect to quinones will be discussed later in this chapter.

Reduction of NAD^+ or $NADP^+$ requires electrons and protons to be transferred to them by an electron transport carrier with a lower redox potential than NAD^+ and $NADP^+$, or for electrons to be pumped up an energy hill using ATP or the high energy intermediate.

This latter process is known as reversed electron transport (see section 2.8.2.)

The mechanism of coupling of ATP synthesis (from ADP and Pi) to electron transport has been a subject of great controversy for many years. The theory now generally accepted is that proposed by Peter Mitchell, for which he received a Nobel prize in 1979. This theory was based on the observation that electron transport mediators were either reduced by electrons or by hydrogen, and that H^+ was transferred across electron transport membranes of mitochondria, bacteria and chloroplasts. Mitchell proposed that an alternative arrangement of electron and hydrogen carriers in an anisotropic (sided) membrane would allow H^+ to be pumped from one side of the membrane to the other as shown in a theoretical 'Mitchell loop' in figure 2.20b. This was then a

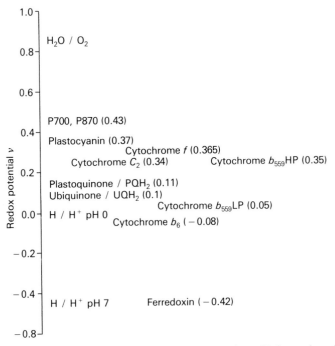

Figure 2.19 A redox potential diagram of some carriers. Unless otherwise stated, values are at pH 7

'proton translocating electron transport chain'. The coupling of this mechanism to ATP synthesis was via a 'proton translocating ATP synthetase' (Figure 2.21). A major supporting experiment for Mitchell's theory was that of Jagendorf who showed that chloroplasts would synthesize ATP following the imposition of a pH gradient (gradient of H^+) across the thylakoid membrane (figure 2.22).

Mitchell proposes that this gradient of protons may conveniently be thought of as two separate energetic contributions

$$\Delta p = \Delta \psi + \Delta pH$$

where Δp is the total proton motive force (pmf), $\Delta \psi$ the electrical potential created across the membrane by movement of H^+ and ΔpH the concentration

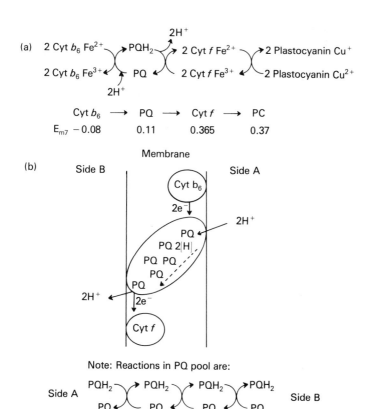

Figure 2.20   (a) Theoretical electron transport chain. (b) Theoretical chemiosmotic (Mitchell) loop. Cyt: cytochrome, PC: plastocyanin, PQ: plastoquinone

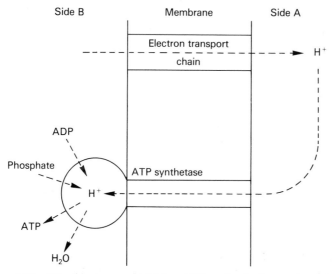

Figure 2.21   Electron transport driving ATP synthesis via a proton gradient

54

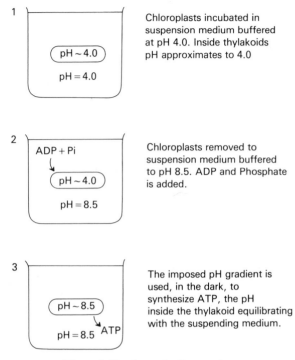

Figure 2.22   Jagendorf's experiment

difference of H^+ between the two sides of the membrane. This latter consideration is important in biological systems as the activity of most enzymes is
dependent upon the pH of the surrounding medium. The control of carbon
fixation in chloroplasts is partly governed by the pH change of the stroma
following the pumping of H^+ into the thylakoid in the light (figure 2.23). This
control mechanism will be discussed in more detail in section 3.10, and more
detail of the ATP synthetase and electron transport chains will be presented
later in this chapter.

2.7.2   Uncoupling of electron transport and ATP synthesis

Electron transport in the absence of ATP synthesis is very slow as the presence
of the proton gradient inhibits further transport of H^+ across the membrane.
ATP synthesis collapses the H^+-gradient, allowing electron transport to proceed. Electron transport and ATP synthesis are said to be coupled. A reagent
which is able to collapse the H^+-gradient is therefore an 'uncoupler' of electron
transport and ATP synthesis. Such a reagent is carbonyl cyanide *p*-
trifluoromethoxyphenylhydrazone (FCCP) which transports H^+ across the
membrane (figure 2.24). Ammonia and amines are also able to collapse the
proton gradient in chloroplasts because of their ability to ionize:

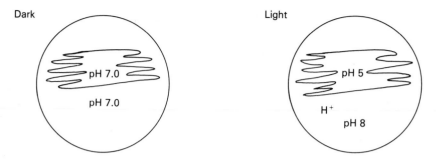

Figure 2.23   The light-induced pH gradient in chloroplasts

$$NH_3 + H^+ \rightleftharpoons NH_4^+$$

$$RNH_2 + H^+ \rightleftharpoons RNH_3^+$$

The membrane is permeable to the uncharged species NH_3 or RNH_2, but impermeable to the charged RNH_3^+ and NH_4^+ (figure 2.25). Thus the ammonia or amine becomes distributed with the pH gradient. Therefore this

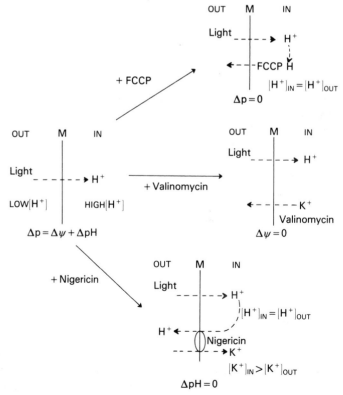

Figure 2.24   The action of FCCP, valinomycin and nigericin

56

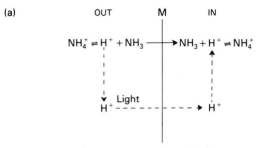

(a)

∴ Each H^+ transported becomes NH_4^+
$\Delta\psi$ is maintained, $\Delta pH = 0$
But $\Delta pH = |NH_4^+|_{IN} - |NH_4^+|_{OUT}$ as all H^+ now NH_4^+

(b) Estimation of ΔpH by using $^{14}CH_3NH_2$, labelled with ^{14}C:

OUT   M   IN

$^{14}CH_3NH_3^+ \rightleftharpoons {}^{14}CH_3NH_2 - - - + - - \rightarrow {}^{14}CH_3NH_2 \rightleftharpoons {}^{14}CH_3NH_3^+$
$+$
$H^+ - - - + - - \rightarrow H^+$
Light

Figure 2.25   (a) Uncoupling by ammonia and (b) the estimation of the pH gradient
across the chloroplast membrane (M)

can be used as a method of estimating the pH gradient across the membrane.
The system to be studied, for example, chloroplasts, is incubated for a short
time with radioactive methylamine ($^{14}CH_3NH_2$), which equilibrates across the
membrane with the pH gradient. Rapid centrifuging separates the membrane
system from the outside solution, and the radioactivity is counted separately in
the precipitate and the supernatant and the pH gradient determined.

Because of the nature of the high energy intermediate, the proton gradient,
its extent and stability is affected by any reagent which alters the permeability
of the membrane to ions. Such reagents are known as ionophores, and three
such compounds have been widely used in photosynthetic research: valinomycin,
nigericin and gramicidin.

Valinomycin transports K^+ across a membrane, and therefore will collapse
the electrical component ($\Delta\psi$) of the protonmotive force, as K^+ will be transferred to equilibrate the charge difference across the membrane. Nigericin will
transport K^+ in exchange for H^+. As a consequence, H^+ is transported from the
side of the membrane with higher concentration to that of lower concentration,
ΔpH being abolished. Addition of both valinomycin and nigericin together
abolishes the protonmotive force, Δp, completely (figure 2.24). Gramicidin
also abolishes Δp by transporting H^+ in the same manner as the uncoupler,
FCCP.

2.8 Photosynthetic electron transfer chains

2.8.1 Photosynthetic bacteria

The simplest photosynthetic electron transfer chains are those of photosynthetic bacteria. The Rhodospirillaceae link the photosynthetic reaction centre to an electron transport chain very similar to that of mitochondria (see figure 2.26). Electrons are transferred from the reaction centre to a cytochromes b/ubiquinone/iron–sulphur centre complex. Reduction of ubiquinone is associated with uptake of H^+ from the aqueous phase, and oxidation with subsequent reappearance of H^+. The electron transfer is cyclic, electrons

Rhodopseudomonas
spheroides

P870 $\longrightarrow$ UQ/Fe $\longrightarrow$ {UQ / Cytochromes b / FeS} $\longrightarrow$ Cytochrome c $\longrightarrow$ Cytochrome c_2 $\longrightarrow$ P870
＿＿＿＿＿＿＿＿＿＿   (bound)
Photosynthetic
reaction centre

Mitochondria

NADH $\longrightarrow$ Fp $\longrightarrow$ Fes $\longrightarrow$ {UQ / Cytochromes b / FeS} $\longrightarrow$ Cytochrome cc_1 $\longrightarrow$ Cytochrome aa_3 $\longrightarrow$ O_2

Figure 2.26   A comparison of the electron transport chain of *Rhodopseudomonas spheroides* (simplified) with that of mitochondria. UQ: ubiquinone, Fp: flavoprotein, FeS: iron sulphur protein

returning to the reaction centre via two cytochromes c. The energy from sunlight produces a charge separation in the reaction centre of approximately 500 mV between the reaction centre chlorophyll, P870 and the acceptor, a ubiquinone associated with an iron atom (FeUQ). The energy release by the reoxidation of FeUQ is used to synthesize ATP. A redox diagram of the chain is shown in figure 2.27.

The reaction centre and light-harvesting complexes of *Rps. spheroides* are the best characterized of any organism. They may be isolated by detergent extraction of the chromatophores and subsequent centrifugation on sucrose gradients and column chromatography. Two light-harvesting complexes are present: LH1 (λ_{max} 875 nm) and LH2 (λ_{max} = 800 nm, 850 nm). These complexes channel energy to a reaction centre containing four bacteriochlorophylls, two bacteriopheophytins, two ubiquinones, one iron atom and some of carotenoids (figure 2.28). The specialized reaction centre chlorophyll is a bacteriochlorophyll dimer, P870, which is oxidized on illumination and passes an electron via a third bacteriochlorophyll and one bacteriopheophytin to a ubiquinone. This ubiquinone is considered to be the primary electron acceptor and is coupled to an iron atom in an unusual way that is not fully understood. The electron finally leaves the reaction centre via a second ubiquinone molecule and enters the complex of two cytochromes b, ubiquinone and an iron–sulphur centre. Electron transfer through this complex is not fully understood and may involve a cycle such as that shown in figure 2.29. This model

58

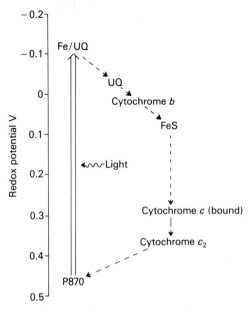

Figure 2.27 A redox diagram of the photosynthetic electron transport chain of *Rhodopseudomonas spheroides*

utilizes the ability of ubiquinone to form radicals, anions and to take up one or two hydrogens. Both the reaction centre and the ubiquinone/cytochrome *b*/iron–sulphur centre complex span the chromatophore membrane, so, as a result of electrons flowing down the electron transport chain, H^+ is transported across the chromatophore membrane (figure 2.30). Thus, as we mentioned earlier in the chapter, Mitchell's requirement that the electron transport chain should act as a proton pump is fulfilled by the bacterial photosynthetic chain. This idea will be discussed in more detail when we have considered the chloroplast electron transport chain (section 2.8.3).

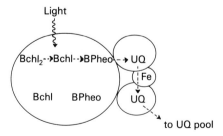

Figure 2.28 A diagrammatic representation of the photosynthetic reaction centre of *Rhodopseudomonas spheroides*. Bchl: bacteriochlorophyll, BPheo: bacteriopheophytin, UQ: ubiquinone

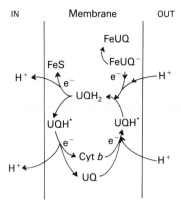

Figure 2.29   A theoretical quinone/cytochrome *b* cycle

2.8.2 *NAD*$^+$ *reduction in photosynthetic bacteria*

The midpoint redox potential of the primary electron acceptor of *Rps. spheroides* ($E_m = -25$ mV at pH 7.0) is too high to enable the organism to reduce NAD$^+$ ($E_m = -340$ mV at pH 7.0). This is true for all purple bacteria, and NAD$^+$ is reduced using the energy of the light-induced proton pump, or of ATP hydrolysis. This energy drives electrons up hill by reversed electron transport from succinate or sulphide to NAD$^+$ (figure 2.31). The green sulphur bacteria are able to reduce NAD$^+$ directly because the midpoint potential of the primary acceptor is sufficiently low ($E_m \sim -550$ mV). The primary acceptor

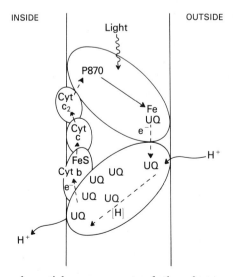

Figure 2.30   Proposed spatial arrangement of the electron transport chain of *Rhodopseudomonas spheroides* in the chromatophore membrane. Cyt: cytochrome, UQ: ubiquinone

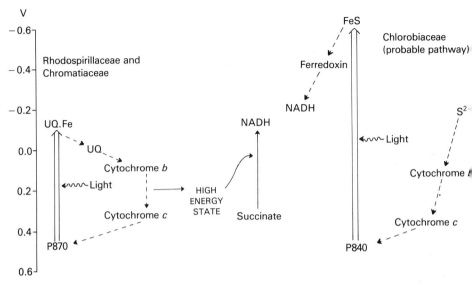

Figure 2.31   NAD⁺ reduction by photosynthetic bacteria

of this organism is an iron–sulphur centre rather than an iron–quinone. Green bacteria use cyclic electron transport to synthesize ATP in the manner described above for purple bacteria.

2.8.3 Chloroplasts and algae (including cyanobacteria)

During the early phases of bacterial evolution in the anaerobic conditions of the primeval world a single highly significant evolutionary step occurred with the appearance of an organism with two photosynthetic reaction centres instead of one. Prokaryotic organisms with this property include the cyanobacteria and the single species of green prokaryotic alga *Prochloron*. One of the photosynthetic reaction centres, photosystem 1, is like that of a green bacterium and is able to reduce NADP directly, having a primary electron acceptor, which is an iron–sulphur centre. The second photosynthetic reaction centre, photosystem 2, shows some structural similarities with that of the purple bacteria, but is able to accept electrons from water rather than H_2S or H_2. This revolutionary step was therefore responsible for the introduction of oxygen into the Earth's atmosphere.

Figure 2.32 shows the generally accepted arrangement of the photosynthetic electron transport chain in algae and chloroplasts with photosystem 1 and photosystem 2 acting in series, connected by electron transport mediators which we will show to be remarkably similar in structure and function to those described in photosynthetic bacteria. Drawing this scheme on a redox scale gives rise to the letter Z, and the scheme is often known as the Z scheme, being originally proposed by Fay Bendall and Robert Hill in 1961. Many investiga-

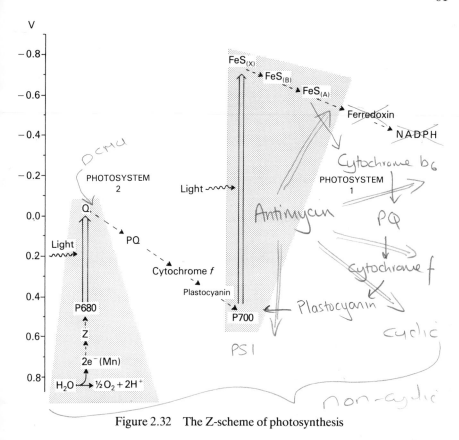

Figure 2.32 The Z-scheme of photosynthesis

tions on this system have used artificial electron donors and acceptors added to isolated thylakoids (figure 2.33). Certain electron transfer carriers may be inhibited by specific inhibitors, in particular, the herbicide dichloromethylurea (DCMU) inhibits electron transfer away from photosystem 2. Robert Hill pioneered the early work in this field, and light-induced oxygen evolution by photosystem 2 linked to an artificial electron acceptor, such as potassium ferricyanide is known as the Hill reaction (see section 1.5).

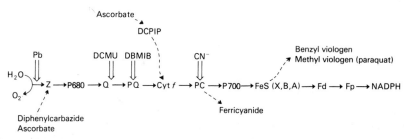

Figure 2.33   Inhibitors of, electron donors to, and electron acceptors from, the photosynthetic electron transport chain. ⇒ inhibitors, ----- electron donors and acceptors

Light-harvesting complexes for photosystems 1 and 2 are not yet as well defined as those associated with bacterial reaction centres. The accessory pigments chlorophyll b (in green algae and plant chloroplasts) phycocyanins and phycoerythrins (in cyanobacteria and red algae) appear to be mainly associated with excitation of photosystem 2, while photosystem 1 is excited by chlorophyll a absorption. The different wavelength dependencies for excitation of photosystems 1 and 2 are a useful experimental tool (see figure 2.34).

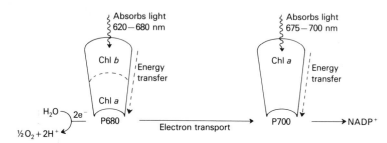

Figure 2.34   Light-harvesting pigments of photosystems 1 and 2 of chloroplasts

2.8.4 Photosystem 2

Photosystem 2 is the least well understood photosynthetic reaction centre. Recent experiments have demonstrated that some similarities may exist on the acceptor side with the reaction centre of photosynthetic bacteria, but the unique donor side, which involves the oxygen evolution mechanism, continues to be a field full of controversy.

· Energy trapped by the light-harvesting complexes is transferred to the specialized reaction centre chlorophyll a, P680 (probably a chlorophyll a dimer) which becomes oxidized, the electron passing via a pheophytin molecule to a primary acceptor, which has been suggested to be a plastoquinone-iron complex similar to the ubiquinone–iron complex found in purple photosynthetic bacteria (see section 2.8.1). This primary acceptor is often known as 'Q' as its identity is not completely verified. Two different compounds absorbing at different wavelengths have been assigned by two groups of workers as being present on the acceptor side of photosystem 2. It may be that the species absorbing at 550 nm (C_{550}) is pheophytin and that 320 nm-absorbing species (X_{320}) is a plastoquinone–iron complex. The reaction centre chlorophyll a, P_{680}, is re-reduced by a donor, D, whose identity is also unknown. This donor is itself re-reduced by electrons derived from water. The splitting of two molecules of H_2O to yield one O_2 releases four electrons, in a four-step mechanism probably involving at least two manganese atoms. Pierre Joliot and Bessel Kok showed that if dark-adapted chloroplasts are illuminated by a series of short flashes of light, then the oxygen evolved per flash was not constant, but showed a periodicity of four (figure 2.35). This led them to the proposal that the oxygen evolving mechanism is capable of five oxidation

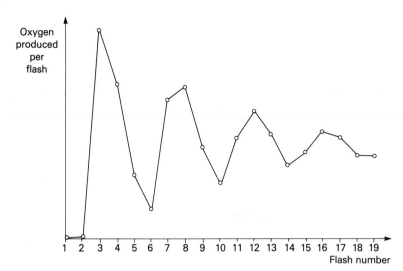

Figure 2.35 Oxygen evolution by chloroplasts excited by a series of short light flashes. (From *Encyclopaedia of Plant Physiology*, volume 5, chapter 8 by B. Dimer and P. Joliot, Springer, New York, 1977)

states, the most oxidizing of which is capable of oxidizing water. If the oxidation states of the system are known as S_0, S_1, S_2, S_3 and S_4, and following each flash of light the excited state of each is denoted by *, then the Kok model is as follows:

$$S_0 \xrightarrow{\text{flash}} S_0{}^*$$
$$S_0{}^* \longrightarrow S_1$$

$$S_1 \xrightarrow{\text{flash}} S_1{}^*$$
$$S_1{}^* \longrightarrow S_2$$

$$S_2 \xrightarrow{\text{flash}} S_2{}^*$$
$$S_2{}^* \longrightarrow S_3$$

$$S_3 \xrightarrow{\text{flash}} S_3{}^*$$
$$S_3{}^* \longrightarrow S_4$$
$$S_4 + 2H_2O \longrightarrow O_2 + 4H^+ + S_0$$

In dark adapted chloroplasts the concentration ratio of these S states is $S_0 : S_1 : S_2 : S_3 = 0.25 : 0.75 : 0 : 0$, which is why the highest yield of oxygen is on the third and fourth flashes. Experiments have shown that manganese is clearly associated with this mechanism but, so far, no models can explain all experimental data.

The majority of chloroplast fluorescence is produced by Photosystem 2, and the behaviour of this fluorescence has been used as a tool to investigate the system. We said earlier (section 2.4) that fluorescence was an overflow system for light energy absorbed by the chlorophylls and transferred to the reaction

64

centre chlorophyll which could not be used to do useful work. A typical experiment is shown in figure 2.36. The normal fluorescence shows a slow rise to a steady state level. The rate of the rise reflects the rate at which the electron transport mediators between the photosystems become reduced. When DCMU is added and chloroplasts illuminated, the rise of fluorescence is much more rapid and the new steady state level is higher as electron transport away from photosystem 2 has been inhibited.

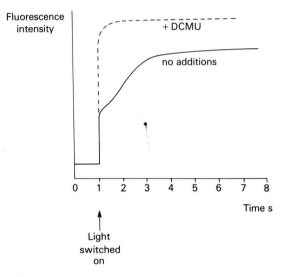

Figure 2.36   Fluorescence emission by photosystem 2

2.8.5 Electron transfer between photosystems 2 and 1

Electrons pass from the primary electron acceptor of photosystem 2, Q, to a pool of plastoquinone which appears to be associated in a complex with a b-cytochrome, cytochrome b_6, an iron–sulphur centre and a c-type cytochrome, cytochrome f (figure 2.36). This complex is probably very similar in its function and behaviour to the complex found in photosynthetic bacteria (see figure 2.30) except that cytochrome b_6 may only take part in cyclic electron transport (see section 2.8.7). Electrons pass from this complex to a copper-containing protein, plastocyanin, and thence are available to re-reduce the photosystem 1 reaction centre chlorophyll, when it has been oxidized in the light.

Apart from cytochromes f and b_6, a third cytochrome, b_{559} is found in chloroplasts. The function of this cytochrome is uncertain as it is difficult to measure the small light-induced absorption changes associated with it. Cytochrome b_{559} appears to exist in two pools of differing redox potential (see figure 2.19) which makes interpretation of available experimental data difficult. The various functions proposed for this cytochrome are as follows:

(1) An electron carrier associated with states S_2 and S_3 of the oxygen evolving system.

(2) As the electron donor, D, to P680 of photosystem 2.

(3) As a carrier on a side path associated with plastoquinone.

(4) As an electron carrier in the electron transport chain between photosystems 2 and 1.

That cytochrome b_{559} is associated with photosystem 2 is very likely as it is found in photosystem 2 preparations isolated from blue-green algae and chloroplasts by detergent extraction.

2.8.6 Photosystem 1

The reaction centre chlorophyll of photosystem 1 shows a characteristic absorption change at 700 nm when oxidized in the light and is therefore called P700. This optical charge was first observed by Bessel Kok. P700 is probably a chlorophyll a dimer, and an electron is passed from this via a third chlorophyll a to an iron–sulphur centre (X) which is the primary electron acceptor. Two other iron–sulphur centres (A and B) are also present in photosystem 1 which accept electrons from centre X. P700 is re-reduced by plastocyanin. The redox potential of centres X, A and B is sufficiently low to reduce NADP directly. Electrons are passed to NADP via two soluble proteins, a flavoprotein enzyme and an iron–sulphur protein called ferredoxin.

2.8.7 Cyclic electron transport

It is possible for electrons to cycle round photosystem 1 in a cyclic electron transport chain (figure 2.37) similar to that seen in photosynthetic bacteria. Electrons return to the cytochrome b_6/PQ/cytochrome f complex in a pathway that can be inhibited by Antimycin A.

2.8.8 The structure of the photosynthetic membrane

The electron and hydrogen carriers and the reaction centres of the chloroplast electron transport chain are ordered in the membrane essentially in the same manner as we have already seen in photosynthetic bacteria. Photosystems 1 and 2 and the plastoquinone/cytochrome/iron–sulphur centre complex all appear to span the thylakoid membrane, as shown diagrammatically in figure 2.38.

2.8.9 Cyclic and non-cyclic phosphorylation

ATP can be synthesized coupled to either non-cyclic electron transport or cyclic electron transport. However, non-cyclic electron transport involves the oxidation of water and the reduction of NADPH, whereas cyclic electron transport only produces ATP. The relative rates of cyclic and non-cyclic

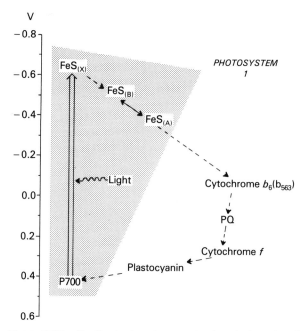

Figure 2.37   Cyclic electron transport using photosystem 1

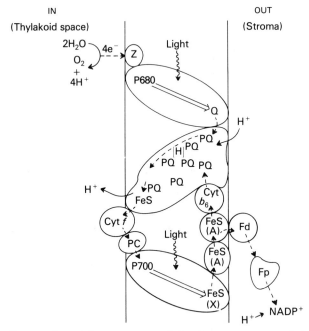

Figure 2.38   Proposed spatial arrangement of the electron transport chain of algae and
chloroplasts in the thylakoid membrane

electron transport are thought to be controlled by the cell's requirements for NADPH and ATP.

2.9 The proton gradient created by electron transport

Much of the early evidence which supported Mitchell's theory of coupling of ATP synthesis to electron transport came from studies on photosynthetic organisms. We have already mentioned Jagendorf's experiment showing that an artificial pH gradient could be used to drive ATP synthesis. An interesting property of photosynthetic organisms can be used to estimate the extent of the electrical gradient formed in the light, that is, the alteration in the pigment spectrum of certain carotenoids in the membrane following the imposition of an electrical gradient across them (electrochromism). The entire spectrum shifts 1 or 2 nm, but the change can be measured conveniently at 515 nm in chloroplasts and 530 nm in *Rps. spheroides*. The extent of the change is directly proportional to the level of $\Delta\psi$ generated in the light. The pools of carotenoids responding to this potential have not been identified in chloroplasts but, in *Rps. spheroides*, these pigments appear to be located in light-harvesting complex 2 (see section 2.8.1). These electrochromic (carotenoid) shifts have been extensively studied by many workers, but the work of Junge and Witt showed that twice the extent of the 515 nm shift was developed when photosystems 1 and 2 were active, as when only photosystem 2 was functioning. This was clear evidence that the two reaction centres produced their charge separation across the membrane.

The light-induced carotenoid shift can be used to calibrate the membrane potential in an experiment such as that developed by Jackson and Crofts which is illustrated in figure 2.39. Addition of valinomycin to chromatophores of *Rps. spheroides* renders the membrane permeable to K^+. Subsequent addition of KCl (in the dark) induces an electrical potential across the membrane as K^+ concentration outside is higher than K^+ concentration inside the chromatophore. This gives rise to a carotenoid shift. By the Nernst equation the electrical membrane potential (the extent of the carotenoid shift) is dependent on the internal and external concentration of K^+ (see figure 2.39). Thus, the carotenoid shift can be calibrated in electrical units such as mV, and the extent of the light-induced shift allows us to calculate the electrical membrane poential developed by the photosynthetic electron transport chain in the light.

By using experimental methods, such as that described above, to measure $\Delta\psi$ and the distribution of radioactive amines to measure ΔpH (see figure 2.25), the amount of energy available for ATP synthesis can be calculated. Table 2.3 shows the data presently available for mitochondria, chloroplasts and photosynthetic bacteria. The contribution of electrical membrane potential ($\Delta\psi$) and the H^+ concentration gradient (ΔpH) to the total protonmotive force clearly varies depending on the system concerned. These differences reflect the differences in composition and ion permeability of the membrane of chloroplasts compared with bacterial chromatophores.

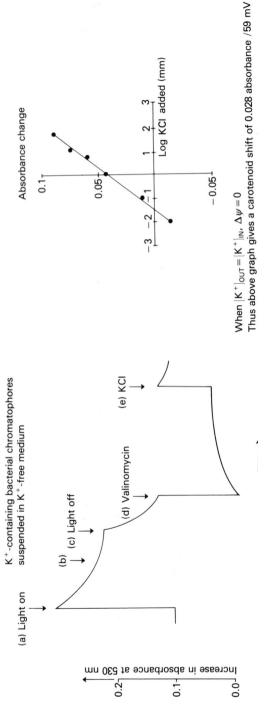

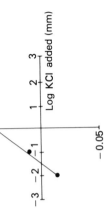

Absorbance change

Log KCl added (mm)

K$^+$-containing bacterial chromatophores suspended in K$^+$-free medium

Increase in absorbance at 530 nm

(a) Light on

(b)

(c) Light off

(d) Valinomycin

(e) KCl

(a) Light on: $\Delta\psi$ formed, shift in carotenoid spectrum seen.
(b) Initial peak of $\Delta\psi$ decreases to steady state level.
(c) Light off: light—induced $\Delta\psi$ decays in dark.
(d) Valinomycin induces diffusion potential negative inside membrane, as $|K^+|_{IN} = |K^+|_{OUT}$
(e) Extra added KCl causes transient $\Delta\psi$ as $|K^+|_{OUT} \neq |K^+|_{IN}$.
  Note: when $|K^+|_{OUT} = |K^+|_{IN}$ no carotenoid shift seen.
  $\Delta\psi = -59 \log \dfrac{|K^+|_{OUT}}{|K^+|_{IN}}$ at 25°C   (Nernst equation)

When $|K^+|_{OUT} = |K^+|_{IN}$, $\Delta\psi = 0$
Thus above graph gives a carotenoid shift of 0.028 absorbance / 59 mV
Light-induced $\Delta\psi$ in experiment above is thus:
Peak = 415 mV
Steady state = 335 mV

Figure 2.39   Estimation of light-induced membrane potential in chromatophores of photosynthetic bacteria using the carotenoid shift

Table 2.3   Comparison of $\Delta\psi$, Δph, Δp generated by different systems

Organelle	$\Delta\psi$ (mv)	ΔpH (59.ΔpH, mV)	Δp (mv)
Mitochondria	160	1.19 (70)	230
Chloroplasts	0*	3.9* (230)	230
Chromatophores (*Rps. spheroides*)	220	1.4 (84)	304

* Measured under high light intensity, steady state.

2.10 ATP synthesis

2.10.1 Stoichiometry of H^+/ATP

The energy available for ATP synthesis is that produced by the protonmotive force, or H^+ gradient, across the bacterial chromatophores or thylakoid membranes. Two questions have been asked by many workers:

(1) How many H^+ are needed for synthesis of 1 ATP?
(2) How many electrons must flow down the electron transport chain in order that one H^+ is transferred across the membrane?

These two questions have proved difficult to answer and much controversy exists over the experimental data available.

One of the problems associated with these measurements is the cyclic nature of the bacterial electron transport chain and the cyclic electron transport around photosystem 1. Using chloroplasts, different workers have obtained values for the ratio of ATP produced for every two electrons passed down the chain from H_2O to NADP that vary between 0.9 and 3.2. Some of this variability may be due to bad chloroplast preparations (at the low end of the range) or the possibility of a contribution from cyclic electron transport (at the high end of the range). Norman Good's research group have investigated the effect of dividing the electron transport chain into different sections by using various inhibitors and artificial electron acceptors, and Trebst's group have measured ATP production separately by photosystems 1 and 2 when the electron transport chain is inhibited between them by DBMIB (see figure 2.33). Both these types of experiments suggest that each photosystem is capable of generating sufficient energy for ATP synthesis, and that two ATP molecules may be formed during electron transfer from H_2O to NADP. The most accurate measurement of the ratio of protons translocated to electrons is to use a pulse of light, move a lump sum of each and count the numbers moved. This can be done by using optical indicators for pH and rapid spectroscopy. The value obtained by these methods is 2.0 H^+ per electron transferred through the chain. For ATP synthesis, reported H^+/ATP ratios vary from 2 to 4. Most recent measurements give values close to 3H^+/ATP which is a little at odds with the ATP to electron flow-rate ratios of 2ATP per 2e⁻. However, thermo-dynamic considerations suggest that 3H⁺/ATP is probably correct. The free energy available from the protonmotive force to do work can be calculated by

$$\Delta G' = 1.36 \, (\Delta pH) + \frac{1.36 \, (\Delta\psi)}{59} \text{ kcal/mole},$$

The $\Delta G'$ for ATP synthesis is -14 kcal/mole, and the predicted values for $\Delta\psi$ and pH if H⁺/ATP = 2 are 342 mV or 5 pH units, respectively. For H⁺/ATP = 3 these values become 208 mV or 3.3 pH units. Reference to table 2.3 will show that $\Delta\psi$ and ΔpH values corresponding to the predictions for H⁺/ATP = 3 are similar to those measured in bacterial chromatophores and chloroplasts.

2.10.2 The ATP synthetase

The mechanism of action of ATP synthetase from mitochondria, chloroplasts and bacterial chromatophores is still under intensive investigation. The proton translocating ATP synthetase is composed of two sectors, a membrane-bound enzyme which performs the forwards and backwards reactions: ATP synthesis and ATP hydrolysis (ATPase)

$$\text{ATP} \underset{\text{ATP synthetase}}{\overset{\text{ATPase}}{\rightleftharpoons}} \text{ADP} + \text{Pi}$$

and a membrane sector which conducts protons across the membrane. The membrane-bound enzyme can be released from the membrane by treatment with EDTA. When released into solution this part of the enzyme only exhibits ATPase activity. Synthesis of ATP only occurs when the membrane-bound component (coupling factor CF_1) is attached to the membrane sector (CF_0) (see figure 2.40). The solubilized CF_1 is assayed by ATPase activity but, unlike the similar enzyme extracted from mitochondria, needs activation by trypsin or heat treatment. Like the mitochondrial coupling factor, chloroplast CF_1 comprises five different polypeptides, designated α, β, γ, δ and ε subunits, in the order of decreasing molecular weight 59 000, 56 000, 37 000, 17 500 and 13 000, respectively. These subunits combine together in a ratio of $2\alpha:2\beta:1\gamma:1\delta:2\varepsilon$ to give an active ATPase enzyme. Nelson has suggested a model for the arrangement of CF_1 when bound to the membrane which is shown in figure 2.40. His experiments suggest that the active site of the enzyme is on the β subunit, giving two active sites per enzyme molecule. The α subunits may be regulatory in function, and it is thought that subunits γ, δ and ε may be concerned with the binding of CF_1 to the thylakoid membrane. This was suggested when an active ATPase lacking subunits γ, δ and ε, unable to bind to the thylakoid membrane, was isolated.

A general model of the proposed mechanism of action is shown in figure 2.41. The molecular mechanisms of this enzyme reaction are still under active investigation.

(a)

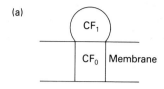

(b)

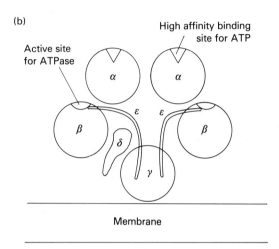

Figure 2.40   The ATP synthetase of chloroplasts. (a) General diagram of the relation-ship between CF_0 and CF_1. (b) Proposed model of the structure of CF_1. (From *Encyclopaedia of Plant Physiology*, volume 5, chapter 7, Chloroplast coupling factors by N. Nelson, Springer, New York, 1977)

2.11 Photosynthetic efficiency

The factors governing the overall efficiency of photosynthesis include the proportion of sunlight that falls on the organism, the efficiency with which that light is harvested and the efficiency of the enzyme reactions involved in electron transfer and carbon and nitrogen fixation.

We can relate the light-harvesting process to photosynthetic rate by measur-ing an action spectrum of photosynthesis, such as that shown in figure 2.42 of

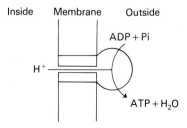

Figure 2.41   General mechanism of ATP synthesis

72

Anacystis nidulans, a blue-green alga. This action spectrum shows the dependence of the rate of oxygen evolution on wavelength. The action spectrum clearly reflects the pigment composition of the organism (see figures 2.6, 2.7). As we have already seen, photosystem 2 is activated by light of shorter wavelength than photosystem 1 (section 2.8.3). In order for oxygen evolution to occur, both photosystems must be activated, so, in order to determine an action spectrum of photosystem 1 alone, a background illumination to active photosystem 2, at 620 nm, must be provided. Similarly, in order to determine an action spectrum for photosystem 2, background illumination to excite photosystem 1 is provided at 690 nm.

These measurements, shown in figure 2.42, confirm the statement that we made earlier with respect to the different pigments associated with light harvesting for the two photosystems. The efficiency of light harvesting and energy transfer to the photosynthetic reaction centres can be measured by the quantum yield of the oxidation of the reaction centre chlorophylls.

$$\text{Quantum yield} = \frac{\text{No. of quanta used to do a specified job}}{\text{No. of quanta absorbed}}$$

When a chlorophyll molecule is excited to a higher energy level, it absorbs one quantum of light of a specific wavelength. The quantum yield of the oxidation of P700 is close to 1, and for photooxidation of P870 (in *Rps. spheroides*) is 1.02 ± 0.04, suggesting that this reaction has 100 % quantum efficiency. The quantum yield of oxygen evolution has been determined by many workers to be ≥8. From the mechanism proposed of four steps in photosystem 2, the S states,

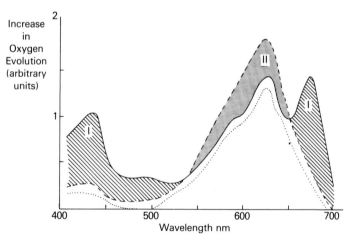

Figure 2.42   Action spectrum of photosynthesis in the blue-green alga *Anacystis nidulars*. ⋯⋯ no background illumination; ------ background illumination at 690 nm (photosystem 1) increases capacity of photosystem 2 shown as shaded area; —— background illumination at 620 nm (photosystem 2) increases capacity of photosystem 1 shown as shaded area. (From *Encyclopaedia of Plant Physiology*, volume 5, chapter 3 by J. D. Radmer and B. Kok, Springer, New York, 1977)

using 4 quanta, and two photosystems in series (photosystem 2 and photo-system 1), 8 quanta (4 + 4) is clearly the minimum possible number.

The overall energy difference generated by photosystems 1 and 2 is about 1 eV. If we calculate the energy of a quantum of light at 700 nm, the longest wavelength that is useful for two photosystem photosynthesis

$$E = \frac{hc}{\lambda} \text{ where } E \text{ is energy}$$

h is Plancks constant
c is speed of light
λ is wavelength

then one quantum at 700 nm possesses energy of 1.8 eV. Thus the highest *energy* efficiency possible is $(1.0/1.8 \times 100)$ % = 40 %. In the photosystems themselves part of this energy is lost in stabilizing the charge separation between the reaction centre chlorophyll and the primary electron acceptor. Figure 2.28 illustrates the proposals made for the reaction centre of *Rps. spheroides*. The greatest reducing capacity generated cannot be used directly, as the probability of the back reaction, of an electron returning from the single Bchl dimer, is very high. The electron must therefore be transferred rapidly away, via this single Bchl and a BPheo to the first stable site, the primary electron acceptor, the iron–quinone. Thus energy is 'wasted' in stabilizing the initial charge separation, i.e., keeping the positive and negative charges apart.

The most direct measure of photosynthetic efficiency *in vivo* is a comparison of the amount of energy obtained upon combustion of the plant material produced. Algae and higher plants have been grown in weak light with ef-ficiencies of about 20 % of the absorbed radiation. This further decrease in efficiency is mainly due to the carbon fixing enzymes.

2.12 Ion movements across the thylakoid membrane and the control of light harvesting in chloroplasts

As we have already seen, when chloroplasts are illuminated, hydrogen ions are pumped into the thylakoid by the action of the electron transport chain (see figure 2.23). In response to this proton gradient, other ions also move across the membrane. Magnesium and calcium ions move out of the thylakoid to the stroma, and chloride ions move into the thylakoid. It appears that Mg^{2+} has some regulatory functions on carbon fixation (see section 3.10), and affects the stacking of thylakoid membranes. It has been shown that the presence of salts in chloroplast isolation medium is essential to maintain thylakoid stacking, the volume of the intrathylakoid space and the thickness of the thylakoid membrane itself.

A further function proposed for magnesium ions is that of regulation of energy distribution between photosystems 1 and 2. Earlier, in section 2.83, we mentioned that light harvesting for the chloroplast was associated with three complexes: a chlorophyll a protein, a chlorophyll b protein and a chlorophyll $a + b$ protein. The first indirect determination of the *size* of the total antennae system of both light reactions goes back to the observation of Emerson and Arnold that one short flash of light gives one O_2 per 2,500 chlorophylls. Since four electrons must flow down the electron transport chain for every O_2 to be evolved, this gave a value of approximately 600 chlorophylls required to harvest energy to excite the two photosystems to move one electron. The existence of different excitation spectra for the two photosystems (section 2.11) suggests that the contributions of pigments to the two photosystems are different. However, there is evidence for 'spill-over' of energy from light-harvesting units serving photosystem 2 to those of photosystem 1. This transfer of energy from photosystem 2 to photosystem 1 is dependent on the concentration of Mg^{2+} and other divalent cations. Butler and Kitajima have proposed a model to show that photosystem 1 and photosystem 2 units may be associated with light-harvesting chlorophylls as shown in figure 2.43. Binding of Mg^{2+} to the

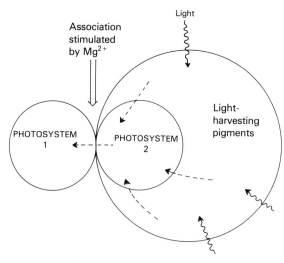

Figure 2.43   Model to show the proposed association between photosystem 1, photosystem 2 and the light-harvesting complexes. ----> Energy transfer

thylakoid membrane may affect the conformation of these closely associated proteins so that the pattern of energy transfer is different. Qualitatively it seems that 'spill-over' of energy from photosystem 2 to photosystem 1 increases the efficiency of the electron transport chain at intermediate light intensities. Under steady state illumination conditions the plastoquinone pool becomes almost completely reduced as photosystem 1 becomes rate-limiting. Under these conditions a H^+-gradient and consequent Mg^{2+}-gradient have been built up across the thylakoid membrane. It is suggested that this Mg^{2+} may function under these conditions to increase energy 'spill-over' to photosystem 1, thus releasing the constraint on electron transport.

2.13 Photosynthesis in Halobacteria

Halobacteria, e.g., *Halobacterium halobium*, are salt tolerant bacteria found in habitats such as salt flats; their optimum growth occurring at 4.3 M sodium chloride concentration, about seven times the normal sea salt concentration. In the presence of high levels of oxygen, halobacteria behave as aerobic bacteria, conserving ATP by respiration. However, under low oxygen levels, purple patches appear on the bacterial membranes which correspond to a unique form of photosynthetic apparatus. Stoeckenius and co-workers isolated this purple membrane fraction and showed that 75 % of the membrane consisted of a single protein of molecular weight 26 000, which bound one mole of retinal per mole of protein. Retinal is also bound as a prosthetic group to visual pigments in the eye, and is capable of a light-induced conformational change shown in figure 2.44, from *cis*-retinal to *trans*-retinal. The protein moiety in halobacteria

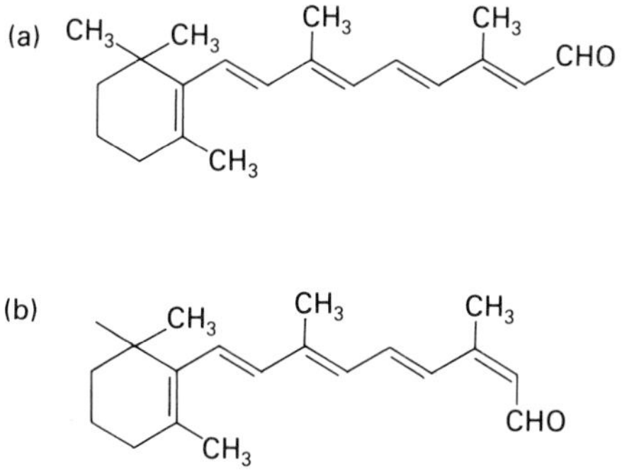

Figure 2.44   Structures of (a) *trans*- and (b) *cis*-retinal

is known as bacterio-opsin, and when bound to retinal the whole is known as bacteriorhodopsin. The linkage between the retinal and opsin is via a Schiff's base, which may be protonated or unprotonated. Illumination causes the dark-adapted bacteriorhodopsin, a mixture of *cis*- and *trans*-retinal isomers absorbing at 560 nm, to become all *trans*, absorbing at 570 nm, in a reaction occurring in $t_{1/2}$ 10 ps. Further illumination causes bleaching of the pigment to a form absorbing at 412 nm. In the latter bleached form the Schiff's base is unprotonated, but protonated in the 570 nm bacteriorhodopsin. The illumination thus results in the loss of H^+ from the bacteriorhodopsin. The protein is situated in such a way in the membrane that protons are pumped from the inside to the outside of the cell on illumination. The H^+-gradient created by this system can be used to drive ATP synthesis in the ATP synthetase complex, similar to that already described (section 2.10). The same ATP synthetase is used for both repiratory phosphorylation and photophosphorylation. The H^+-gradient can also be used as an energy source to drive the active transport of ions and amino acids in halobacteria.

Suggested Further Reading

General articles and books

Barber, J. (Ed.) (1977). *Topics in Photosynthesis*, vol. 2, *Primary Processes of Photosynthesis*, Elsevier, Amsterdam

Clayton, R. K. and Sistrom, W. R. (Eds.) (1978). *The Photosynthetic Bacteria*, Plenum Press, New York

Miller, K. R. (1979). The photosynthetic membrane, *Sci. Amer.*, October

Trebst, A. and Avron, M. (Eds.) (1977). *Encyclopaedia of Plant Physiology*, new series, vol. 5, *Photosynthesis I – Photosynthetic Electron Transport and Phosphorylation*, Springer, New York

Reviews on electron transport and electron transport carriers

Amesz, J. and van Gorkom, H. J. (1978). Delayed fluorescence in photosynthesis, *Ann. Rev. Plant Physiol.*, **29**, 47–66

Blankenship, R. E. and Parson, W. W. (1978). The photochemical electron transfer reactions of photosynthetic bacteria and plants, *Ann. Rev. Biochem.*, **47**, 635–654

Butler, W. L. (1978). Energy distribution in the photochemical apparatus of photosynthesis, *Ann. Rev. Plant Physiol.*, **29**, 345–378

Cramer, W. A. and Whitmarsh, J. (1977). Photosynthetic cytochromes, *Ann. Rev. Plant Physiol.*, **28**, 133–278

Govindjee and Braun, B. Z. (1974). Light absorption, emission and photosynthesis, in W. D. P. Stewart (Ed.), *Algal Physiology and Biochemistry*, Blackwell Scientific, Oxford, pp. 346–390

Hall, D. O., Cammack, R., Rao, K. K., Evans, M. C. W. and Mullinger, R. (1975). Ferredoxins, blue-green bacteria and evolution, *Biochem. Soc. Trans.*, **3**, 361–368

Levine, R. P. (1974). Mutant studies on photosynthetic electron transport, in W. D. P. Stewart (Ed.) *Algal Physiology and Biochemistry*, Blackwell Scientific, Oxford, pp. 424–433

Malkin, R. (1982). Photosystem 1, *Ann. Rev. Plant Physiol.*, **33**, 455–578

Velthuys, B. R. (1980). Mechanisms of electron flow in photosystem 2 and towards photosystem 1, *Ann. Rev. Plant Physiol.*, **31**, 545–567

Reviews on the high energy state and phosphorylation

Hall, D. O. (1976). The coupling of electron transport to phosphorylation in isolated chloroplasts in J. Barber (Ed.), *The Intact Chloroplast*, Elsevier, Amsterdam, pp. 135–170

Hauska, G. and Trebst, A. (1977). Proton translocation in chloroplasts, *Curr. Top. Bioenerg.*, **6**, 152–221

Hinkle, P. C. and McCarty, R. E. (1978). How cells make ATP, *Scientific American*, March

Junge, W. (1977). Membrane potentials in photosynthesis, *Ann. Rev. Plant Physiol.*, **28**, 503–536

McCarty, R. E. (1979). Roles of coupling factor for phosphorylation in chloroplasts, *Ann. Rev. Plant Physiol.*, **30**, 79–104

Shavit, N. (1980). Energy transduction in chloroplasts: structure and function of the ATPase complex, *Ann. Rev. Biochem.*, **49**, 111–138

Reviews on regulation by ion movements

Barber, J. (1976). Ionic regulation in intact chloroplasts and its effect on primary photosynthetic processes, in J. Barber (Ed.), *The Intact Chloroplast*, Elsevier, Amsterdam, pp. 89–134

Buchanan, B. B. (1980). Role of light in the regulation of chloroplast enzymes, *Ann. Rev. Plant Physiol.*, **31**, 341–74

Reviews of general interest

Hall, D. O. (1979). Solar energy use through biology – past, present and future, *Solar Energy*, **22**, 307–328

Olson, J. M. (1978). Precambrian evolution of photosynthetic and respiratory organisms, in Max K. Hecht, W. C. Steere and B. Wallace (Eds.), *Evolutionary Biology*, vol. 11, Plenum Press, New York, pp. 1–37

Experimental methods

Holton, D. and Windsor, M. W. (1978). Picosecond flash photolysis in biology and biophysics, *Ann. Rev. Biophys. Bioeng.*, **7**, 180–227

Methods in Enzymology, **23A**, Photosynthesis A (Ed. A. san Pietro) (1971)

Methods in Enzymology, **69**, Photosynthesis and Nitrogen Fixation C (Ed. A. san Pietro) (1980)

3

Metabolism in Photosynthetic Systems

3.1 Introduction

In chapter 2 the phosphorylation of ADP to ATP in photosynthetic membranes was discussed: the so-called 'light reactions' of photosynthesis. Reducing equivalents (NADH and NADPH) are also formed. This chapter considers how this ATP and NAD(P)H are used by enzymes in the cytosol of photosynthetic bacteria and the stroma of chloroplasts to reduce carbon dioxide and synthesize carbohydrate. For convenience glucose is presumed to be the product of this CO_2 fixation, but the final end products are polysaccharides, lipids and amino acids, and their biosynthesis will also be discussed (section 3.6).

This CO_2 fixation or reduction of CO_2, often referred to as the 'dark reactions', is only associated with photosynthetic systems. There are certain reactions catalysed by biotin-containing enzymes in animals which fix CO_2, e.g., pyruvate carboxylase in mitochondria and acetyl Coenzyme A carboxylase in the cytosol, but, in both cases, the CO_2 that is fixed by these enzymes is lost in a subsequent reaction. The overall reactions are shown briefly in figure 3.1. Other reactions involving the release of CO_2, like that catalysed by isocitrate dehydrogenase in the TCA cycle, are essentially irreversible and therefore there are no reactions in animals capable of fixing CO_2 and leading to a net gain of carbohydrate.

Early attempts to isolate chloroplasts, in the middle of this century, produced chloroplasts that had lost their outer membrane and stroma. They could phosphorylate ADP to produce ATP, but they were unable to fix CO_2. This implied that the enzymes for CO_2 fixation could be in the outer membrane or stroma of chloroplasts. When methods were later devised to isolate intact chloroplasts (see section 1.6) these were able to fix CO_2 at rates similar to that in leaves, and it is now well established that the majority of the enzymes are in the stroma.

Full accounts of the enzymes of the Calvin cycle, the 'dark reactions', including the history of their discovery and experimental evidence for their existence, can be found in most general biochemistry text books. In this chapter only a brief description of the Calvin cycle is given, greater emphasis being given to the control and organization of this metabolic pathway and others associated with photosynthetic systems (section 3.10).

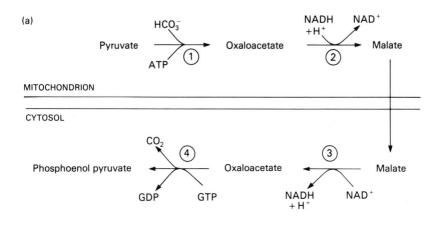

(a)

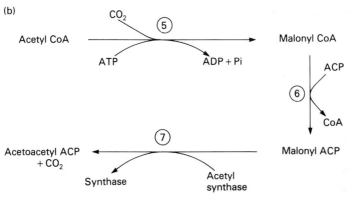

(b)

Figure 3.1 (a), (b) Reactions catalysed by carboxylases: (1) pyruvate carboxylase (EC.6.4.1.1), (2) malate dehydrogenase (EC.1.1.1.37), (3) malate dehydrogenase, (4) PEP carboxykinase (EC.4.1.1.32), (5) acetyl CoA carboxylase (EC.6.4.1.2), (6) ACP malonyl transferase (EC.2.3.1.39), (7) 3-oxoacyl ACP synthase (EC.2.3.1.41)

3.2 The Calvin cycle

CO_2 fixation into carbohydrate in eukaryotic photosynthetic membranes is often referred to as the Calvin, Bassham and Benson cycle after the researchers who elucidated the enzymic reactions. In this book it will be referred to as the Calvin cycle. It is also called the 'dark reactions' of photosynthesis because, if there is a plentiful added supply of ATP and NAD(P)H (reducing equivalents), which are formed in the 'light reactions' (see section 2.7), chloroplasts will fix some CO_2 into carbohydrate. However, photophosphorylation and the Calvin cycle are intimately linked and it is only for convenience that they are discussed separately as the 'light and dark reactions'. In fact, light plays an important role in the control of some of the enzyme reactions in the Calvin cycle (see section 3.10).

80

The Calvin cycle, as it occurs in chloroplasts, is represented in a simplified form in figure 3.2, emphasizing the cyclic nature of this metabolic pathway. Briefly, 6 carbons in the form of CO_2 react with 6 molecules of ribulose 1,5 bisphosphate to form 12 molecules of triose phosphate, and 2 of these give rise to glucose ($C_6H_{12}O_6$) which is taken out of the cycle. The dotted arrow indicates the interconversions that the other 10 triose phosphate molecules undergo, forming 4,5,6 and 7 carbon sugar phosphates and eventually bringing

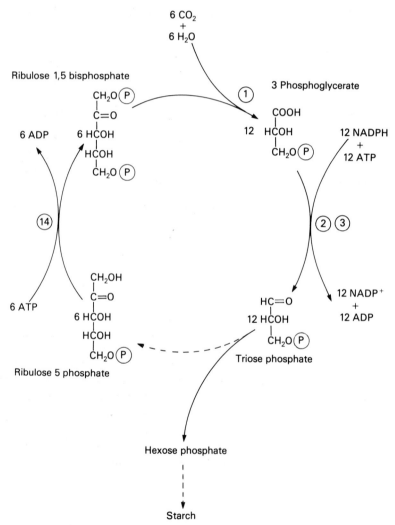

Figure 3.2   The Calvin cycle. Dotted lines represent several enzyme catalysed reactions: (1) ribulose bisphosphate carboxylase (RBPC) (EC.4.1.1.39), (2) phosphoglycerate kinase, (3) glyceraldehyde 3 phosphate dehydrogenase, (14) phosphoribulose kinase, ⓟ phosphate group

about the regeneration of 6 molecules of ribulose 1,5 bisphosphate, allowing the cycle to continue.

3.2.1 Ribulose 1,5 bisphosphate carboxylase

Ribulose 1,5 bisphosphate carboxylase (also referred to in textbooks as ribulose diphosphate carboxylase) is considered as the first enzyme of the Calvin cycle, as it reacts with CO_2. The enzyme catalyses a two-step reaction, where the intermediate 2 carboxy 3 ketoribitol 1,5 diphosphate is formed and then hydrolysed (figure 3.3). The enzyme has a high affinity for CO_2 ($K_m \sim 0.15$

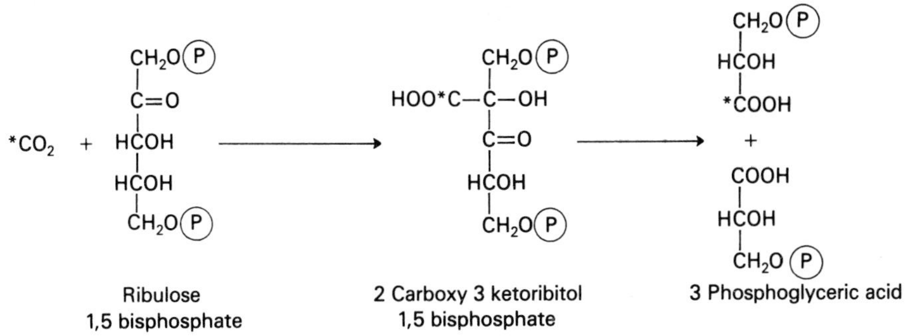

Figure 3.3   The reaction catalysed by ribulose bisphosphate carboxylase. *Carbon from radiolabelled CO_2

mM), while that for bicarbonate is low ($K_m \sim 22$ mM). The products of the reaction are 2 molecules of 3 phosphoglyceric acid (3PGA) per molecule of ribulose bisphosphate. The asterisk in figure 3.3 indicates the fate of the carbon from radiolabelled CO_2, showing that initially only 50 % of the 3 PGA would become labelled.

The carboxylase enzyme is present in the chloroplast in very large quantities; it represents 15 % of the total chloroplast protein and it may be bound to the outer surface of the thylakoid membrane. The enzyme has a very large diameter (20 nm) and a molecular weight of 550 000 has been determined. This is made up from 16 subunits: 8 subunits are catalytic, each of molecular weight 55 000 and the other 8 subunits are regulatory, each of molecular weight 13 000. The enzyme is allosteric and controlled by the level of oxygen, light intensity, pH and various metabolites. It is the main controlling point of the Calvin cycle and its regulation will be discussed in more detail in sections 3.5 and 3.10.

3.2.2 The other reactions of the cycle

The next two reactions in the Calvin cycle bring about the reduction of 3 phosphoglycerate to glyceraldehyde 3 phosphate (figure 3.4 (1) and (2)); these reactions are analogous to those occurring in gluconeogenesis. The

3 phosphoglycerate is phosphorylated by the enzyme phosphoglycerate kinase to form 1,3 diphosphoglycerate. The phosphate is transferred from ATP to 3 phosphoglycerate. The 1,3 diphosphoglycerate is then reduced to glyceraldehyde 3 phosphate by the action of glyceraldehyde 3 phosphate dehydrogenase, using the coenzyme NADPH, unlike the related enzyme in glycolysis and gluconeogenesis which is NAD^+-linked. These two reactions represent points in the cycle where ATP and reducing equivalents are required.

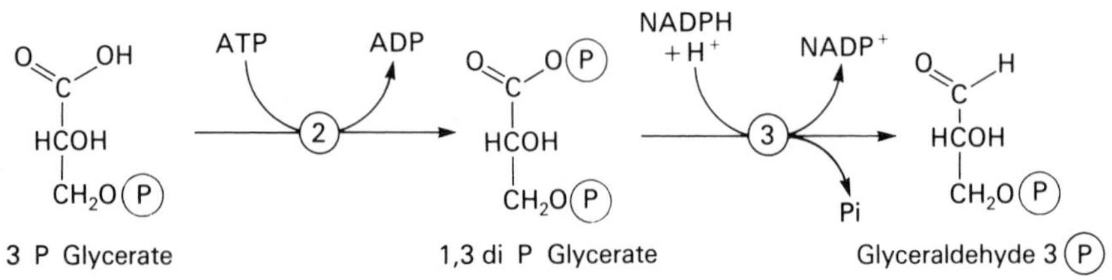

Figure 3.4   Reactions catalysed by phosphoglycerate kinase (EC.2.7.2.3) and glyceraldehyde 3 phosphate dehydrogenase (EC.1.2.1.9)

The Calvin cycle is shown in detail in a linear form in figure 3.5. Here 2 molecules of glyceraldehyde 3 phosphate are being used to form 1 molecule of glucose by enzymes which also occur in gluconeogenesis, i.e., phosphoglycero isomerase, adolase, fructose diphosphatase, phosphohexose isomerase and glucose 6 phosphatase. The other 10 molecules of glyceraldehyde 3 phosphate undergo a series of interconversions which bring about the regeneration of ribulose 1,5 bisphosphate, thus giving the pathway its cyclic form. A major enzyme involved in these interconversions is transketolase. This is a thiamine pyrophosphate (TPP)-containing enzyme and in the presence of Mg^{2+} it catalyses the transfer of a glycolaldehyde group from one sugar phosphate to another (figure 3.6). The TPP binds the glycolaldehyde group (CH$_2$OH.CO-) from fructose 6 phosphate and sedoheptulose 7 phosphate (reaction (9), figure 3.5), and adds it on to glyceraldehyde 3 phosphate to give xylulose 5 phosphate.

The result of all these changes is that from 10 molecules of glyceraldehyde 3 phosphate, 4 of xylulose 5 phosphate and 2 of ribose 5 phosphate are formed. These are all converted to ribulose 5 phosphate by the action of ribose phosphate isomerase and ribulose phosphate epimerase, respectively. These reactions involve the isomerization of certain hydroxyl groups. The ribulose 5 phosphate is then phosphorylated by phosphoribulose kinase to give ribulose 1,5 bisphosphate, the phosphate being donated by ATP. This is therefore another reaction in which the ATP synthesized by photophosphorylation is used (section 2.10) and the substrate for ribulose bisphosphate carboxylase is regenerated.

Some of the reactions of the Calvin cycle are the same as those occurring in the pentose phosphate pathway in the cytosol. The reactions catalysed by transketolase occur in both metabolic pathways. However, some of the reactions do not occur in the pentose phosphate pathway, e.g., those catalysed by

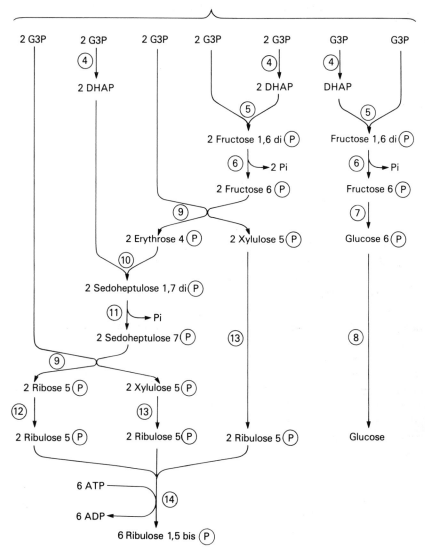

Figure 3.5   The Calvin cycle. G3P: glyceraldehyde 3 phosphate, DHAP: dihydroxy-acetone phosphate. (1) RBPC, (2) and (3) see figure 3.4, (4) triose phosphate isomerase (EC.5.3.1.1), (5) fructose diphosphate aldolase (EC.4.1.2.13), (6) fructose diphosphatase (EC.3.1.3.11), (7) phosphohexose isomerase (EC.5.3.1.9), (8) glucose 6 phosphatase (EC.3.1.3.9), (9) transketolase (EC.2.2.1.1), (10) sedoheptulose diphosphate aldolase (EC.2.2.1.2), (11) sedoheptulose diphosphatase (EC.3.1.3.37), (12) ribose phosphate isomerase (EC.5.3.1.6), (13) ribulose phosphate 3 epimerase (EC. 5.1.3.1), (14) phosphoribulose kinase (EC.2.7.1.19)

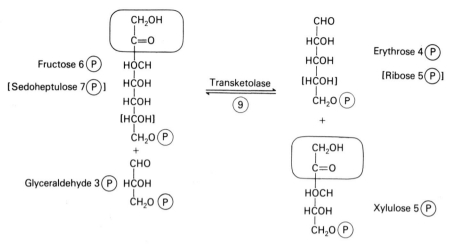

Figure 3.6   The reaction catalysed by transketolase: the transfer of a glycolaldehyde group ($CH_2OH.CO-$)

sedoheptulose diphosphate phosphatase and fructose diphosphatase, so the Calvin cycle is not a complete reversal of the pentose phosphate pathway.

3.2.3 Energy requirements

In the Calvin cycle, for each CO_2 fixed into carbohydrate, i.e., for 1 turn of the cycle, 3 ATP and 2 NADPH molecules are required. Of the ATP molecules 2 are used in the reaction catalysed by phosphoglycerate kinase (reaction 2 in figure 3.4) since two molecules of 3 phosphoglycerate are formed when 1 CO_2 is fixed and 2 NADPH molecules are used in the reaction catalysed by glyceraldehyde phosphate dehydrogenase (reaction 3). The third ATP molecule is used to phosphorylate ribulose 5 phosphate (reaction 14 in figure 3.5). The overall reaction for the biosynthesis of 1 molecule of glucose is as follows:

6 Ribulose bisphosphate + 18 ATP + 12 NADPH + 12H$^+$ + 6CO$_2$ + 12H$_2$O →
$C_6H_{12}O_6$ + 6 Ribulose bisphosphate + 18 ADP + 18 Pi + 12 NADP$^+$
  glucose

Since there is no net gain or loss of ribulose bisphosphate it can be cancelled out of the overall equation.

$$6CO_2 + 18\,ATP + 12\,NADPH + 12\,H^+ + 12\,H_2O \rightarrow$$
$$Glucose + 18\,ADP + 18\,Pi + 12\,NADP^+$$

This overall equation indicates how the hydrolysis of ATP (ATP + H_2O → ADP + Pi) is used to drive biosynthetic reactions. The hydrolysis of ATP does occur, but not as represented in the equation above since it is used to phosphorylate metabolites in reactions 2 and 14 (figure 3.5) with the formation of ADP. The inorganic phosphate (Pi) is hydrolysed from different metabolites at reactions 3,6,8 and 11.

3.3 CO_2 fixation in C_4 plants

In the 1960s it was discovered that in certain plants the metabolites which were labelled initially, when radiolabelled CO_2 ($^{14}CO_2$) was supplied, which contain 4 carbons and not 3 phosphoglycerate (a C_3 acid), were C_4 acids, e.g., oxalo-acetate, malate and aspartate. The pathway by which C_4 acids were labelled first is known as the Hatch and Slack pathway after the researchers who elucidated it. The plants that carry out this type of CO_2 fixation are characteristically mainly tropical and very efficient at producing hexose at high light intensities and high temperatures. They grow rapidly due to their fast rate of hexose biosynthesis. The rate of CO_2 fixation on a leaf area basis is 2 or 3 times faster in C_4 plants than C_3 plants, which only use the Calvin cycle. These plants have what is called a Kranz-type anatomy. There are two concentric layers of cells around the vascular bundles, the 'bundle sheath' and a sheet forming part of the mesophyll. Both types of cells have chloroplasts, but the chloroplasts of the bundle sheath have very few grana (figure 1.8) and have a low level of photosystem 2 (section 1.5.2). Maize and sugar cane are examples of C_4 plants. The Hatch and Slack pathway is shown in figure 3.7.

In these C_4 plants the CO_2 is fixed in the cytosol of the mesophyll cells by the enzyme phosphoenol pyruvate (PEP) carboxylase ((1) in figure 3.7) and the oxaloacetate so formed is reduced to malate by $NADP^+$-linked malate de-hydrogenase. Malate is subsequently transported into the chloroplasts of the bundle sheath cells and, via the action of malic enzyme, CO_2 is regenerated in the bundle sheath cells' chloroplasts where it is used by the Calvin cycle as described in section 3.2. Pyruvate is also generated and this is transported back into the mesophyll cells where the action of pyruvate phosphodikinase regen-erates PEP. As PEP carboxylase in the mesophyll cells has a higher affinity for CO_2 than ribulose bisphosphate carboxylase in the bundle sheath cells, this mechanism has the effect of concentrating the CO_2 in the bundle sheath cells for the Calvin cycle. This concentrating effect probably accounts for the higher rates of photosynthesis in C_4 plants than C_3 plants, because it enables the RBPC enzyme, which has a high K_m, to function more efficiently.

This pathway requires more energy than the Calvin cycle alone. Pyruvate phosphodikinase transfers phosphate from ATP to pyruvate to form PEP and AMP. In order to regenerate ADP from AMP, another ATP is required in the adenylate kinase (EC.2.7.4.3) reaction.

$$AMP + ATP \rightleftharpoons 2ADP$$

Therefore, for 6 CO_2 molecules to be fixed into glucose an extra 6×2 ATP molecules are required making a total of 30. This cycle has another effect, and that is the effective transfer of reducing equivalents (NADPH) from the mesophyll cells to the bundle sheath cells by the action of malate de-

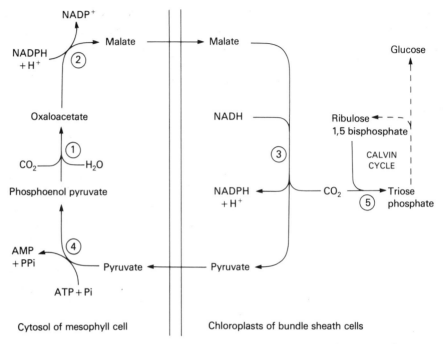

Figure 3.7 CO_2 fixation in C_4 plants: (1) PEP carboxylase (EC.4.1.1.13), (2) NADP-malate dehydrogenase (EC.1.1.1.82), (3) Malic enzyme (EC.1.1.1.40), (4) pyruvate diphospho kinase (EC.2.7.9.1), (5) RBPC

hydrogenase, then malic enzyme (see figure 3.7). Since there is only a low level of photosystem 2 in the chloroplasts of the bundle sheath cells, little NADPH will be produced there, so this transfer tops up the amount required for CO_2 fixation.

The pathway as shown in figure 3.7 shows the formation of malate from oxaloacetate, but it has been proposed that there are two sorts of C_4 plants, the 'malate formers' and the 'aspartate formers'. The mechanism for the formation of aspartate, however, is not clear. Aspartate is formed by the transamination of oxaloacetate with glutamate.

$$\text{Oxaloacetate} + \text{Glutamate} \rightleftharpoons \text{Aspartate} + \text{Oxoglutarate}$$

Then possibly aspartate is transferred into the bundle sheath cells where the reverse reaction regenerates oxaloacetate, which can be decarboxylated to give pyruvate and CO_2. The plants which form aspartate as an early reaction in the pathway, the 'aspartate formers', do have high levels of aspartate amino transperase.

CO_2 fixation in C_4 plants is more efficient than in C_3 plants and it has been proposed that this is due partly to the lack of photorespiration in C_4 plants (section 3.5). Photorespiration is when oxygen is used by RBPC instead of CO_2

and the enzyme is then referred to as an oxygenase. CO_2 fixation by RBPC is inhibited by oxygen, which is then metabolized to remove the inhibition. In C_4 plants RBPC is in the bundle sheath cells, where the oxygen concentration is low and therefore there is little or no photorespiration.

The control of CO_2 fixation in C_4 plants is discussed in section 3.10.2.

3.4 Crassulacean acid metabolism (CAM)

CO_2 fixation, where the first product is a C_4 metabolite, also takes place in another group of plants. It is called crassulacean acid metabolism (CAM) after one family of plants which uses this metabolic pathway to fix CO_2. The acid formed is malate or malic acid (COOH. CHOH. CH$_2$. COOH) and this type of C_4 metabolism occurs in many succulent plants, e.g., Crassulaceae, Cactaceae (cacti), Liliaceae (lilies) and Orchidaceae (orchids) and others. The characteristics of these plants are that they are slow growing and are found in very hot, dry climates. During the day it is so hot that the stomata have to remain closed to prevent water loss. Therefore when light is available for photosynthesis, the CO_2 may not be able to enter the cells. It was observed that in these plants the concentration of malate increased overnight, and decreased during the day.

During the night the CO_2 is fixed into oxaloacetate by a reaction catalysed by PEP carboxylase (figure 3.8).

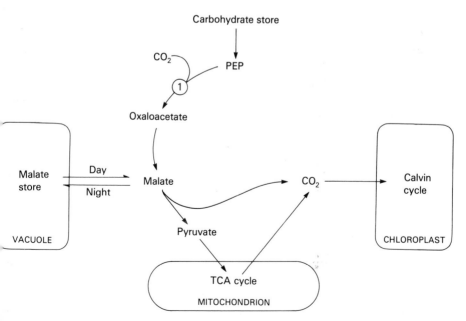

Figure 3.8   Crassulacean acid metabolism. (1) PEP carboxylase

$$\text{Phosphoenol pyruvate} + CO_2 \rightleftharpoons \text{Oxaloacetate} + Pi$$

The oxaloacetate is converted to malate by NAD^+-linked malate dehydrogenase. Both these enzymes are present in the cytosol of the leaf cells. The PEP for this reaction is derived from carbohydrate stores through glycolysis or by the action of ribulose bisphosphate carboxylase, since both can give rise to triose phosphates.

The malate is transported from the cytosol into the cell vacuoles, where it is stored. Concentrations of 100–200 mM malate may be achieved in the vacuoles; in these plants the size of the vacuoles is relatively greater as a percentage of cell size than in plants which do not operate CAM.

The malate acts as a store of CO_2 within the plant. These reactions occur at night and the rates are increased by a decrease in temperature, which would occur at night. The malate is probably transported into the vacuole by an active transport system, since such high concentrations of malate are achieved. PEP carboxylase is inhibited by malate, so the compartmentalization of malate in the vacuole also allows this enzyme to remain active and therefore CO_2 fixation can continue.

During the day (figure 3.8) the malate is mobilized and transported back into the cytosol where pyruvate and CO_2 are formed by the action of malic enzyme.

$$\text{Malate} + NADP^+ \rightleftharpoons \text{Pyruvate} + CO_2 + NADPH + H^+$$

The CO_2 is then fixed by the Calvin cycle (section 3.2) in the chloroplasts and pyruvate is probably transferred into the mitochondria, where it is metabolized by the tricarboxylic acid cycle.

3.5 Photorespiration

Respiration is the process by which metabolites are oxidized by molecular oxygen. This occurs in mitochondria where NADH and CO_2 are produced by the tricarboxylic acid cycle and NADH is then reoxidized by an enzyme in the electron transport chain. The electrons obtained from this oxidation are passed by the chain to reduce oxygen to give H_2O. Plant cells have mitochondria which carry out this 'dark' respiration, consuming oxygen and producing carbon dioxide. Coupled to the transfer of electrons from NADH to oxygen is the phosphorylation of ADP to give ATP (see also section 2.10). This is the cell's most efficient way of producing energy. However, in photosynthetic systems there is another oxygen-consuming system where metabolites are oxidized and CO_2 is released: this is called photorespiration. In this case ATP and reducing equivalents (NAD(P)H) are required.

In plant cells mitochondrial respiration is apparently inhibited by light, at least, the inhibitors of electron transport, rotenone and antimycin A, have no effect in the light. In addition, it is not affected to any extent by the oxygen concentration since it is saturated at oxygen concentrations of about 4 %. Photorespiration requires light and its rate is proportional to the oxygen concentration. Photorespiration is initiated when the oxygen level increases in the presence of light because ribulose bisphosphate carboxylase is inhibited by oxygen, but it is able to use oxygen as a substrate. When oxygen is the substrate instead of CO_2 the enzyme is referred to as ribulose bisphosphate oxygenase. The reactions of photorespiration are shown in figure 3.9. The products of the oxygenase reaction in the chloroplast are 3 phosphoglycerate and phosphoglycolate. Glycolate, formed by the hydrolysis of phosphoglycolate, is transported into the peroxisomes where it is oxidized by the flavin enzyme glycolate oxidase to glyoxylate. During this reaction oxygen is consumed and hydrogen peroxide is formed which is metabolized by catalase (figure 3.10). Further metabolic reactions bring about the formation of intermediates of the Calvin cycle and ultimately ribulose bisphosphate, the substrate for RBPC, is reformed.

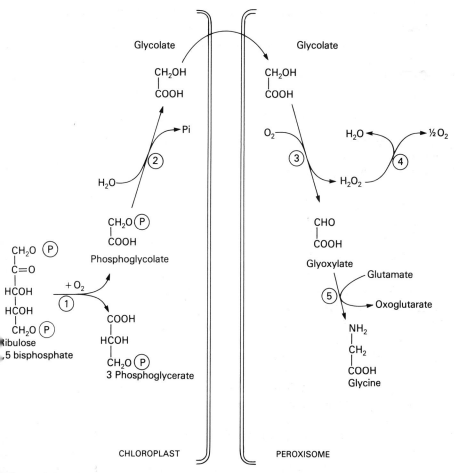

Figure 3.9 Photorespiration: (1) RBP oxygenase, (2) phosphoglycolate phosphatase (EC.3.1.3.18), (3) glycolate oxidase (EC.1.1.3.1), (4) catalase (EC.1.11.1.6), (5) glycine aminotransferase (EC.2.6.1.4)

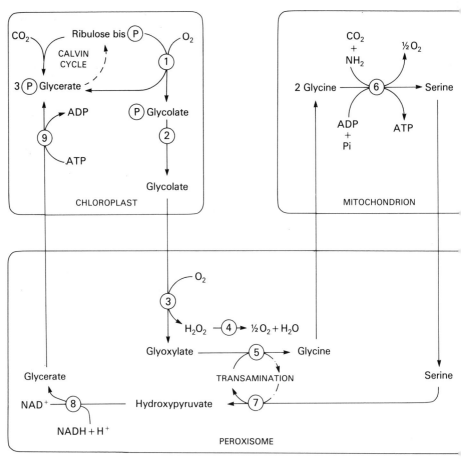

Figure 3.10 Photorespiration showing the possible fate of glyoxylate and the release of CO_2. (1) to (5) see figure 3.9, (6) reaction linked to oxidative phosphorylation in mitochondria, (7) serine aminotransferase (EC.2.6.1.45), (8) glycerate dehydrogenase (EC.1.1.1.29), (9) glycerate kinase (EC.2.7.1.31)

The glyoxylate undergoes transamination (see figure 3.13) with glutamate to give glycine and oxoglutarate. The glycine is transported into the mitochondria, where serine is synthesized by the oxidation of glycine, giving rise to ATP and CO_2. The serine can be transaminated to give hydroxypyruvate which is converted to glycerate by a NADH-linked dehydrogenase. The enzymes for the last two reactions are present in the peroxisomes. The glycerate is phosphorylated and the triose phosphate sugar produced can give rise to intermediates of the Calvin cycle in the chloroplast, or be used for the biosynthesis of carbohydrates etc. (section 3.6). Most of these reactions require a large energy input in the form of ATP or NAD(P)H.

The overall effect of photorespiration is to reduce the oxygen concentration and therefore relieve the inhibition of ribulose bisphosphate carboxylase. It has also been suggested that photorespiration acts to transport ATP or to transfer fixed carbon out of the chloroplasts. However, there is a transport mechanism in the chloroplast membrane for transporting ATP (see section 3.9), and photorespiration would be very inefficient, since for 1 ATP formed during photorespiration a large number of ATPs and reducing equivalents are required. There is also a transporting system for triose phosphates in the chloroplast membrane, allowing the movement of fixed carbon from the Calvin cycle into the rest of the cell.

Research is currently being carried out on the effects of inhibition of photorespiration in C_3 plants, as this could increase crop yields in these plants. Photorespiration is inhibited by lowering the $O_2 : CO_2$ ratio; crop yields can be nearly doubled by a five-fold increase in CO_2 concentration. Obviously this can only be achieved in a closed environment; crop yields of greenhouse grown tomatoes are increased by releasing CO_2 into the atmosphere.

C_4 plants do not exhibit photorespiration. This may be because the Calvin cycle, and therefore ribulose bisphosphate carboxylase, is in the bundle sheath cells (the inner layer of cells) in these plants. The CO_2 is concentrated there and the oxygen concentration is low, and so photorespiration is unlikely to operate. However, C_4 plants do contain peroxisomes and it is thought that when the stomata are closed in these plants in dry conditions, the CO_2 concentration would be too low for the Calvin cycle to operate and phosphoglycolate would be formed by the action of RBP oxygenase. The CO_2 then formed by photorespiration, using the peroxisome and mitochondrial enzymes, is then fixed by the Calvin cycle.

3.6 The formation of polysaccharides, fatty acids and amino acids

The Calvin cycle is represented in section 3.2 as the biosynthesis of glucose by enzymes in the stroma of the chloroplast. However, glucose is not the only product of the Calvin cycle and a large number of polysaccharides, fatty acids and amino acids are formed as a result of CO_2 fixation. The products are often referred to collectively as photosynthates.

The end products of CO_2 fixation are not necessarily formed in the chloroplast. For instance, one of the main end products of photosynthesis in higher plants is sucrose, i.e., glucose and fructose linked by an $\alpha1-2$ glycosidic linkage (see figure 3.11), but sucrose cannot be transported across the chloroplast inner membrane, therefore it is reasonable to assume that sucrose is synthesized outside the chloroplast. It is generally thought that triose phosphates, e.g., phosphoglyceric acid and dihydroxyacetone phosphate (DHAP), are the end products of the Calvin cycle in the chloroplast and these are transported into the cytosol, where further metabolism takes place. The transport of metabolites will be discussed in more detail in section 3.9.

Sucrose is synthesized by the enzyme sucrose synthetase in the cytosol from glucose and fructose (see figure 3.11). These are both derived from DHAP, which is produced by the Calvin cycle. Glucose in the form of glucose 6

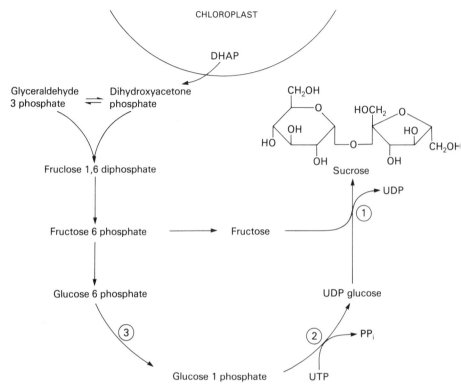

Figure 3.11 Biosynthesis and structure of sucrose: (1) sucrose synthetase (EC.2.4.1.13), (2) glucose 1 phosphate uridyl transferase (EC.2.7.7.9), (3) phosphoglucomutase (EC.2.7.5.1)

phosphate undergoes isomerization to give glucose 1 phosphate, which reacts with UTP to give UDP-glucose and pyrophosphate. The UDP glucose reacts with fructose to give sucrose; this is the reaction catalysed by sucrose synthetase. ATP and GTP can be used instead of UTP, but the reaction is much slower. Sucrose is readily transported throughout the plant via the phloem. Sucrose synthetase is a reversible enzyme, allowing sucrose to be synthesized at a site of rapid Calvin cycle activity and then transported to another part of the plant, where hexose (glucose) can be reformed from the sucrose.

Starch, a major polysaccharide end product of CO_2 fixation, is formed in the chloroplast (see the starch granules in figure 1.2 and 1.8) as well as in the cytosol. Starch occurs in two forms, α amylose and amylopectin (see figure 3.12). Amylopectin is a highly branched polysaccharide consisting of glucose residues linked by an α 1–4 glycosidic bond, with branch points of α 1–6 glycosidic bonds. It is similar to the storage polysaccharide, glycogen. However, amylopectin has branches 24–30 monosaccharides long, while glycogen has 8–12 monosaccharides in each branch. α amylose consists of a long unbranched chain of glucose residues linked by α 1–4 glycosidic bonds. Starch is

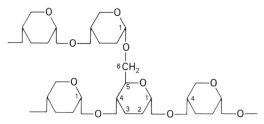

Figure 3.12   Structure of starch. Amylopectin is highly branched with D-glucose residues linked by α 1–4 and α 1–6 glycosidic linkages. Amylose consists of D-glucose residues linked by α 1–4 glycosidic bonds forming a long unbranched molecule which forms a helix

formed by the action of starch synthetase which, like glycogen synthetase, adds a monosaccharide, in the form of ADP-glucose, to some existing polysaccharide.

$$\text{ADP-glucose} + (\text{Glucose})_n \rightarrow (\text{Glucose})_{(n + 1)} + \text{ADP}$$
$$\text{starch}$$

This is a similar reaction to that catalysed by sucrose synthetase (figure 3.11). Starch is the main store of carbohydrate in plants and, when required, can be broken down to give glucose. Fats are stored in the form of triacyl glycerols, i.e., fatty acids bound to glycerol, these are also called triglycerides. Fatty acids $(CH_3(CH_2)_nCOOH)$ are synthesized in plants from acetyl Coenzyme A (acetyl CoA) by the multi-enzyme complex, fatty acid synthetase. The action of this enzyme complex is the same as in animals. Acetyl CoA is formed from triose phosphates from the Calvin cycle, by the action of enzymes from glycolysis and pyruvate dehydrogenase. The last reaction takes place in the mitochondria and chloroplasts, and therefore the acetyl group must be transported into the cytosol for fatty acid synthesis to take place. It has been suggested that acetyl CoA can cross the plant mitochondrial membrane, but the acetyl group may well be transported as citrate or attached to carnitine as in mammalian mitochondria. If acetate is supplied, acetyl CoA can be formed by the action of acetyl CoA synthetase in the cytosol as follows:

$$\text{Acetate} + \text{ATP} + \text{CoA} \xrightarrow{\text{Mg}^{2+}} \text{Acetyl CoA} + \text{AMP} + \text{PPi}$$

Saturated fatty acids are formed by the action of fatty acid synthetase, but a large number of unsaturated fatty acids are required by plants, e.g., in membranes (see section 1.4.1). The formation of these unsaturated fatty acids involves a large number of enzyme reactions and will not be discussed here but can be found in other textbooks.

For amino acid, protein and nucleic acid biosynthesis, nitrogen as well as carbon is required in a useable form. The nitrogen in the air (dinitrogen, $N \equiv N$) is very inert and cannot be used directly by plants. However, atmospheric nitrogen can be used by various microorganisms to form ammonia

(NH_3) and nitrate (NO_3^-), which can be assimilated into the amino group $(-NH_2)$ of glutamate. This will be discussed in section 3.8.

The amino group from glutamate can be transferred to different keto acids by transamination (see figure 3.13) to make other amino acids. Two amino acids, which are formed rapidly by transamination of glutamate with oxaloacetate and pyruvate, are aspartate and alanine. The carbon skeletons for these two amino acids are derived from 3 phosphoglycerate (see figure 3.14). The 3 phosphoglycerate from the Calvin cycle is converted to phosphoenol pyruvate

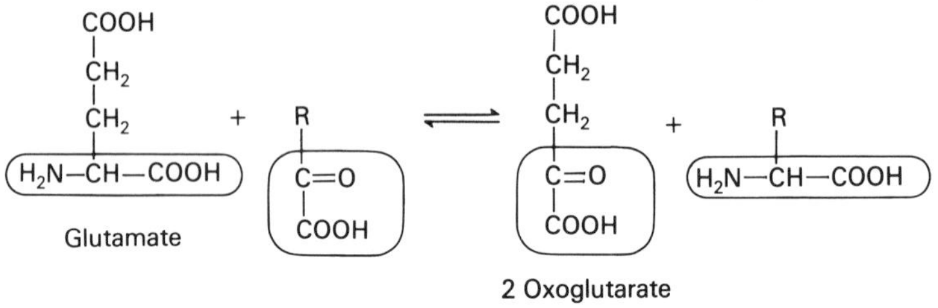

COOH
|
CH₂
|
CH₂   +   R   ⇌   CH₂   +   R
| | | |
H₂N—CH—COOH   C=O   C=O   H₂N—CH—COOH
| | | |
Glutamate   COOH   COOH

2 Oxoglutarate

Figure 3.13   A transamination reaction. R represents the amino acid side chain. The trivial name of transaminases often comes from the amino acid with the R side chain, e.g., if R = −H the enzyme is called glycine aminotransferase

(PEP) by the action of phosphoglyceromutase and phosphopyruvate hydratase in the cytosol. The pyruvate and oxaloacetate are formed by the action of pyruvate kinase and PEP carboxylase, respectively, on PEP (see figure 3.14).

Glycine and serine are also synthesized rapidly in photosynthetic systems, and this involves the cooperation of peroxisomes and mitochondria (see section 3.5 on photorespiration). Glycine is formed by the transamination of glyoxylate with glutamate in peroxisomes (see figure 3.9).

Glyoxylate + Glutamate ⇌ Glycine + Oxoglutarate

Serine is derived from glycine in mitochondria by a reaction coupled to the synthesis of ATP (see figure 3.10). A description in detail of the biosynthesis of all the amino acids could take a whole textbook and, therefore, will not be given here. Briefly, glutamate is a precursor of arginine and glutamine, aspartate is a precursor of threonine, lysine and isoleucine, and pyruvate is a precursor of valine and leucine. All these reactions occur in bacteria, fungi and plants and illustrate how the carbon skeletons of amino acids can be derived from intermediates of the Calvin cycle. The aromatic amino acids are derived from the compound shikimic acid (figure 3.15). The condensation of PEP and erythrose 4 phosphate is the first step in the biosynthesis of shikimic acid. Both of these compounds come from the Calvin cycle. Shikimic acid is very important in bacteria and higher plants, because as well as being the precursor of tyrosine, phenyl-alanine and trytophan, it is also a precursor of lignin (which occurs in the woody portion of plants) and ubiquinone and plastoquinone (which are involved in electron transfer – see section 2.6.1).

The biosynthesis of the nucleotides for nucleic acid synthesis (RNA and

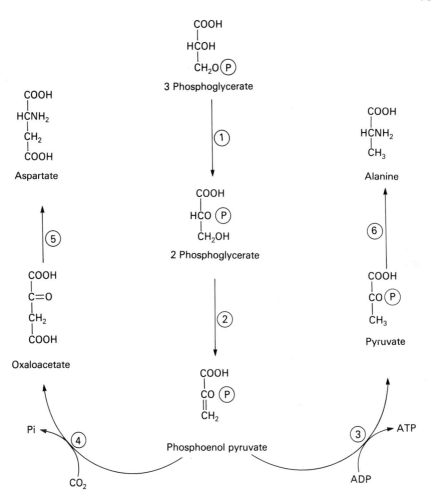

Figure 3.14 Formation of aspartate and alanine from 3 phosphoglycerate: (1) phosphoglyceromutase, (2) phosphopyruvate hydratase (EC.4.2.1.11), (3) pyruvate kinase (EC.2.7.1.40), (4) PEP carboxylase, (5) aspartate aminotransferase (EC.2.6.1.1), (6) alanine aminotransferase (EC.2.6.1.2)

Figure 3.15 The formation of shikimic acid

DNA) involves a large number of enzymic reactions. Briefly, the purine and pyrimidine rings are derived from certain amino acids, e.g., aspartate and glycine, and the ribose is derived from ribose 5 phosphate, which is an intermediate of the Calvin cycle.

3.6.1 Glyoxylate cycle

The glyoxylate cycle is another method by which plants and microorganisms can synthesize carbohydrate. In the tricarboxylic acid cycle in mitochondria there are two CO_2 releasing reactions catalysed by isocitrate dehydrogenase and oxoglutarate dehydrogenase. Oxaloacetate from the TCA cycle is used for the biosynthesis of glucose (gluconeogenesis). A 2-carbon molecule of acetate (acetyl CoA) condenses with oxaloacetate to form citrate at the start of the TCA cycle and 2 CO_2 molecules are released during the TCA cycle, so there can be no *net* synthesis of glucose from acetate. In plants and microorganisms there can be net synthesis of glucose from acetate by the action of the glyoxylate cycle. The two enzymes of the cycle which do not occur in the TCA cycle are isocitrate lyase and malate synthase. These are present in the glyoxysomes. The whole cycle is shown in figure 3.16. It has been found that the enzymes malate dehydrogenase and citrate synthase are present in the glyoxysomes and therefore the whole cycle could operate there. Usually, however, isocitrate, the substrate for the lyase, is believed to be transported from mitochondria to the glyoxysomes and the products, succinate and malate, are transported back from the glyoxysomes to the mitochondria.

3.7 CO_2 fixation in bacteria

There are several metabolic pathways for the assimilation of carbon in bacteria, and which pathway is used depends on the species and the growing conditions. Some bacteria can survive under anaerobic or aerobic conditions, e.g., *Rs. rubrum*, and with a variety of carbon sources, e.g., CO_2 or acetate. This is often exploited in the laboratory, where growing conditions can be kept constant for fixed lengths of time, and then altered, and the resulting metabolic changes observed. This section gives brief summary of the possible mechanisms of carbon fixation present in photosynthetic bacteria; this is more correctly called photoassimilation of carbon by bacteria.

The overall purpose of this photoassimilation of carbon in bacteria is slightly different to that in plants and algae. Bacteria do store some carbohydrate (glycogen or polyhydroxybutyrate), but they require rapid production of metabolites for new proteins and lipids, etc., for cell reproduction and development. Therefore, although the Calvin cycle operates in most bacteria, additional carboxylation reactions are present (see figure 3.17). These include the glyoxylate cycle (section 3.6.1), the TCA cycle operating in reverse and individual reactions, e.g., phosphoenol pyruvate (PEP) and pyruvate carboxylases. Many bacteria contain ribulose bisphosphate carboxylase (RBPC, see

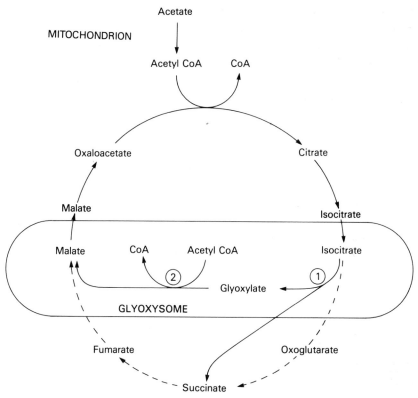

Figure 3.16   The glyoxylate cycle: (1) isocitrate lyase (EC.4.1.3.1), (2) malate synthase (EC.4.1.3.2)

section 3.2.1) and its presence has been used to indicate the presence of CO_2 fixation. This enzyme is controlled in photosynthetic bacteria in a similar way to that in plant chloroplasts (see section 3.10). However, there are three possible molecular sizes of RBPC present in photosynthetic bacteria; large (molecular weight $\sim$ 500 000), intermediate (molecular weight $\sim$ 360 000) and small (molecular weight $\sim$ 120 000). The large enzyme occurs mainly in purple sulphur bacteria, e.g., *Chromatium vinosum*, the intermediate enzyme occurs in green bacteria, e.g., *Rhodopseudomonas palustris*, and the small enzyme occurs in *Rs. rubrum*. The large enzyme is similar to that found in plants and contains 8 small and 8 large subunits (section 3.2.1). The intermediate and small enzymes contain 6 and 2 large subunits, respectively, and no small subunits. There is evidence that RBPC can act as an oxygenase as well as carboxylase in bacteria containing all three sizes of enzyme. In plants the small subunit has a regulatory role. Presumably the mechanism of regulation in the intermediate and small enzymes must be different from that where the small subunit is present. The control of RBPC is discussed in section 3.10.

The glyoxylate cycle generally operates when acetate is the sole source of

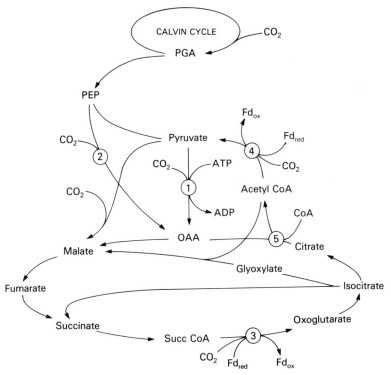

Figure 3.17 Carboxylation reactions associated with the Calvin cycle in photosynthetic bacteria: (1) pyruvate carboxylase, (2) PEP carboxylase, (3) oxoglutarate synthase, (4) pyruvate synthase, (5) citrate lyase

carbon for a particular bacterium. Under these circumstances the Calvin cycle enzymes would be repressed. In *Chromatium*, malate and oxoglutarate dehydrogenases are absent. These are two important enzymes involved in the synthesis of ATP in mitochondria and bacteria, where they operate as TCA cycle enzymes generating NADH for the electron transport chain. The ATP needs of *Chromatium* are presumably fulfilled by the light reactions, and the assimilation of carbon can be accomplished by either the Calvin cycle or the glyoxylate cycle, depending on whether the organism is growing on CO_2 or acetate.

The *Rhodospirillaceae* have been used extensively for the study of bacterial carbon assimilation. Generally these organisms use the Calvin cycle. Some, e.g., *Rhodospirillum rubrum*, have a complete TCA cycle, which allows them to grow aerobically in the dark. *Rs. Rubrum* does not have a glyoxylate cycle or PEP carboxylase. *Rhodopseudomonas palustris* has a complete TCA cycle and also a glyoxylate cycle and PEP carboxylase (figure 3.17). The metabolic pathways present in these related bacteria within the same family indicate how carbon assimilation can vary even between members of the same taxonomic group.

There are certain reactions in the TCA cycle where CO_2 is released, e.g., those catalysed by pyruvate and oxoglutarate dehydrogenases. These reactions are essentially irreversible, and the action of pyruvate dehydrogenase leads to a net loss of carbohydrate stores. Some photosynthetic bacteria can apparently fix CO_2 by a reversal of these reactions, i.e., a reversal of the TCA cycle. They use the powerful reducing agent ferredoxin to carry out the reductive carboxylation of acetyl and succinyl CoA, which NADH cannot do (see figure 3.18). Reduced ferredoxin is produced by the 'light reactions' (see section 3.2) and the reactions are catalysed by pyruvate and oxoglutarate synthases. The cycle, where the TCA cycle is effectively reversed, is called the reductive carboxylic acid cycle and it operates in some of the Chlorobiaceae (green bacteria), where the Calvin cycle enzyme activities appear to be very low. However, there is conflicting evidence about the role of this cycle, since it requires other key enzymes, e.g., citrate lyase, which cannot be found in some species of Chlorobiaceae.

(a)   PYRUVATE SYNTHASE

Acetyl CoA + CO_2 + Ferredoxin$_{red}$ → Pyruvate + CoA + Ferredoxin$_{ox}$

(b)   OXOGLUTARATE SYNTHASE

Succinyl CoA + CO_2 + Ferredoxin$_{red}$ → Oxoglutarate + CoA + Ferredoxin$_{ox}$

Figure 3.18   Reactions catalysed by (a) pyruvate synthase (EC.1.2.7.1) and (b) oxoglutarate synthase (EC.1.2.7.3)

There are several mechanisms for photoassimilation of carbon in photosynthetic bacteria. The metabolic pathway generally present is the Calvin cycle, but other ancillary mechanisms for carbon fixation are present, e.g., PEP carboxylase and the glyoxylate cycle, which assume greater importance over the Calvin cycle under certain growing conditions.

3.8 Nitrogen metabolism

The biosynthesis of amino acids and purine and pyrimidine bases, for proteins and nucleic acids, requires reduced nitrogen in the form of $-NH_2$. There is an abundance of nitrogen in the air (81 %), but this is in the extremely inert form of dinitrogen (N ≡ N). Higher plants cannot utilize this nitrogen, but certain microorganisms can reduce nitrogen, to ammonia (NH_3) and this is called nitrogen fixation. Other microorganisms can convert this ammonia to nitrite and nitrate, e.g., *Nitrosomonas* and *Nitrobacter*, respectively. Higher plants take up the nitrogen through their roots in the form of nitrate (NO_3^-). Plants then use the enzymes nitrate and nitrite reductases to reform ammonia, which is used for the biosynthesis of glutamate and hence other amino acids and nucleotides (section 3.6). This 'nitrogen cycle' is illustrated in figure 1.16.

3.8.1 Nitrogen fixation

The first step in the nitrogen cycle is nitrogen fixation, the reduction of N_2 to NH_3. This is carried out by certain prokaryotes, e.g., photosynthetic bacteria, cyanobacteria and aerobic soil bacteria (*Azotobacter*). There are also symbiotic bacteria, e.g., *Rhizobium*, which are found in the nodules on the roots of leguminous plants (peas and beans), where they can fix nitrogen in symbiosis with the plants.

The reduction of N_2 is carried out by an enzyme complex called nitrogenase. The enzyme consists of two proteins. The first is called the molybdenum–iron protein (MFP) and is a tetramer (2×2 different subunits) of molecular weight 220 000, containing 2 atoms of molybdenum and about 25 atoms of non-haem iron and acid labile sulphide. The second protein is called iron protein (FP) and is a dimer of molecular weight 60 000 containing 4 iron atoms and acid labile sulphide groups. These two proteins are very similar, whatever the source of the enzyme. The Mo–Fe protein from one organism can mix with iron protein from another organism and give an active hybrid enzyme.

A major problem in the study of nitrogenase is that it is inactivated by oxygen. This means that all isolation and kinetic work on the enzyme, including such techniques as electrophoresis and centrifugation, has to be carried out in the absence of air. This inactivation of nitrogenase by oxygen also poses a problem for the nitrogen fixing organisms. Anaerobic conditions for the nitrogenase can be achieved in several ways. If the organism is an obligate anaerobe, e.g., *Clostridium pasteurianum*, or can function anaerobically (a facultative anaerobe), e.g., *Rs. rubrum*, then nitrogen fixation will only occur under anaerobic conditions.

Under nitrogen fixing conditions, i.e., when NO_3^- or NH_3 is not available, certain aerobes, e.g., cyanobacteria (blue-green algae) such as *Nostocaceae* and *Anabena sp.*, differentiate certain cells in the filament to form heterocysts. In these specialized nitrogen fixing cells, photosystem 2 is repressed, therefore no oxygen is evolved and no CO_2 fixed. The energy for nitrogen fixation therefore has to be imported from the adjacent cells. The symbiotic bacteria are able to keep oxygen away from the nitrogenase by the action of leghaemoglobin. This is synthesized by the plant and binds oxygen in a similar way to haemoglobin. It also accounts for the pink colour of the root nodules.

The nitrogenase enzyme is often assayed using acetylene ($CH \equiv CH$) as substrate; this is reduced to ethylene ($CH_2 = CH_2$) which can be assayed rapidly using gas chromatography. It is possible to measure nitrogen fixation in a great many systems, e.g., nodules, soil samples or isolated enzyme, using this method. The enzyme requires a reducing agent, energy (ATP) and Mg^{2+} which is probably bound to the ATP. The reducing agent is usually reduced ferredoxin, which is produced in the light reactions. A possible overall reaction is as follows:

$$N_2 + 3XH_2 + 12\,ATP \rightarrow 2\,NH_3 + 3X + 12\,ADP + 12Pi$$

The number of ATP molecules may vary from organism to organism, but a large amount of ATP is required, which cannot be used for cell growth. X indicates the reducing agent, which could be ferredoxin or NADH; for assays in the laboratory sodium dithionite is used. A schematic representation of the reaction is shown in figure 3.19.

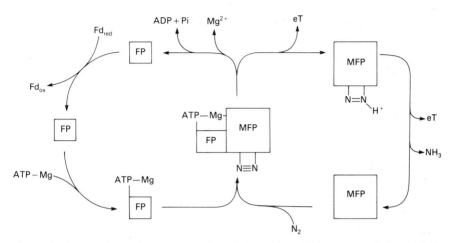

Figure 3.19   A schematic representation of the action of nitrogenase (EC.1.18.2.1). MFP: molybdenum–iron protein, FP: iron protein, eT: electron transfer, Fd: ferredoxin

All aspects of the nitrogenase reaction are under intense investigation. One line of investigation is to isolate the genes that code for the enzyme and see if they might be incorporated into another organism (genetic engineering). Only legumes are able to fix nitrogen for their own needs, because of the symbiotic bacteria in their root nodules. Other food plants, e.g., corn, are supplied with fertilizers. These contain fixed nitrogen (NH_3 or NO_3^-), which is produced chemically and requires a large amount of energy in its production (Haber process). Environmental problems are now arising because excess nitrates are washed into rivers and lakes. Lake-dwelling organisms such as cyanobacteria (blue-green algae), which would normally use large amounts of ATP for nitrogen fixation, because of the presence of NO_3^-, use the ATP for increased growth instead. This problem of excessive growth (eutrophication) threatens the normal balance of organisms in lakes. There is therefore a great need to find ways of exploiting the nitrogenase enzyme to save fuel, energy and pollution. The control of the enzyme will be discussed in section 3.10.

3.8.2 Nitrate reduction

Plants take up nitrogen in the form of nitrate. Organisms such as *Nitrosomonas* and *Nitrobacter* convert ammonia from the action of nitrogenase to nitrite and then nitrate respectively in the soil. However, the plants require nitrogen in the form of ammonia for amino acid biosynthesis, so the nitrate taken up by the roots has to be converted to ammonia. This is carried out by the enzymes nitrate and nitrite reductases.

In plant cells, nitrate reductase is in the cytosol, although there is evidence that it is loosely bound to the chloroplast envelope. It is a flavoprotein containing FAD, molybdenum, iron and sulphur. NADH from respiration is the probable source of reducing power. A simplified representation of the reduction of nitrate is shown in figure 3.20a. The reaction catalysed by nitrite reductase is shown in figure 3.20b and this occurs in the chloroplasts. Electrons are accepted by the enzyme from reduced ferredoxin generated during photosynthetic electron flow.

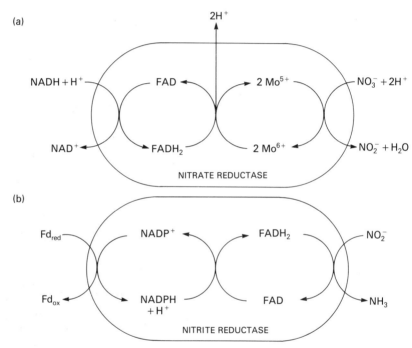

Figure 3.20   The action of (a) nitrate reductase (EC.1.6.6.4) and (b) nitrite reductase (EC.1.6.6.3). Mo: molybdenum

These two enzymes also occur in some bacteria, e.g., cyanobacteria. In these bacteria they are not in separate compartments, but are membrane bound and their action is intimately linked to photosynthesis.

3.8.3 Assimilation of ammonia

The ammonia produced in prokaryotes and higher plants has to be converted to the amino group of glutamate. Other amino acids can be formed by the transamination of glutamate with a keto acid (section 3.6). There are two enzyme reactions, which bring about the formation of the amino group of glutamate from ammonia. The reaction shown in figure 3.21a is catalysed by glutamate dehydrogenase and in figure 3.21b the reactions are catalysed by glutamine and glutamate synthetases. Both mechanisms require reducing equivalents and glutamate synthetase requires ATP. Glutamate dehydrogenase has a low affinity for ammonia (high $K_m \sim 2$ mM), while glutamine synthetase has a high affinity for ammonia. ($K_m \sim 0.2$ mM). Therefore the glutamate dehydrogenase reaction will only occur when the concentration of NH_3 is high. Since NH_3 is an uncoupler (see section 2.7.2) in chloroplasts, it is desirable that its concentration is kept low, and so the glutamine/glutamate synthetase pathway for glutamate synthesis is probably more likely to occur in higher plants.

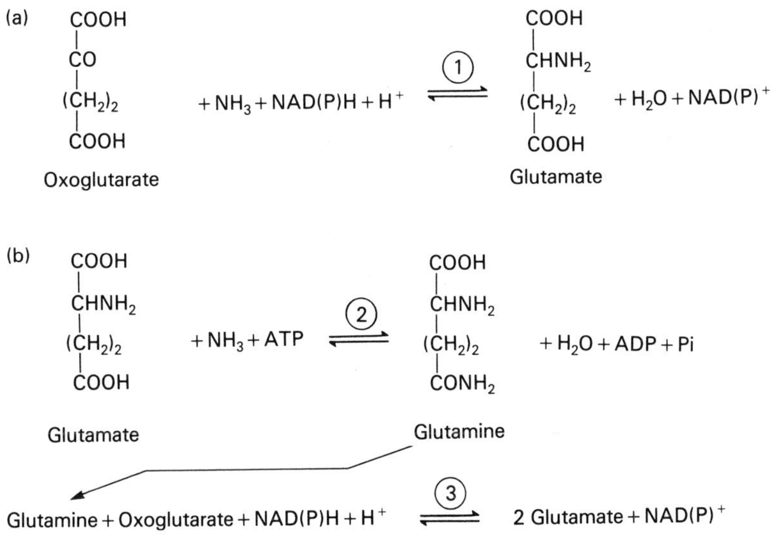

(a)

$$\underset{\text{Oxoglutarate}}{\overset{\displaystyle \text{COOH}}{\underset{\displaystyle \text{COOH}}{\overset{\displaystyle |}{\underset{\displaystyle |}{\overset{\displaystyle \text{CO}}{\underset{\displaystyle (\text{CH}_2)_2}{|}}}}}}} + NH_3 + NAD(P)H + H^+ \;\underset{1}{\rightleftharpoons}\; \underset{\text{Glutamate}}{\overset{\displaystyle \text{COOH}}{\underset{\displaystyle \text{COOH}}{\overset{\displaystyle |}{\underset{\displaystyle |}{\overset{\displaystyle \text{CHNH}_2}{\underset{\displaystyle (\text{CH}_2)_2}{|}}}}}}} + H_2O + NAD(P)^+$$

(b)

$$\underset{\text{Glutamate}}{\overset{\displaystyle \text{COOH}}{\underset{\displaystyle \text{COOH}}{\overset{\displaystyle |}{\underset{\displaystyle |}{\overset{\displaystyle \text{CHNH}_2}{\underset{\displaystyle (\text{CH}_2)_2}{|}}}}}}} + NH_3 + ATP \;\underset{2}{\rightleftharpoons}\; \underset{\text{Glutamine}}{\overset{\displaystyle \text{COOH}}{\underset{\displaystyle \text{CONH}_2}{\overset{\displaystyle |}{\underset{\displaystyle |}{\overset{\displaystyle \text{CHNH}_2}{\underset{\displaystyle (\text{CH}_2)_2}{|}}}}}}} + H_2O + ADP + Pi$$

Glutamine + Oxoglutarate + NAD(P)H + H$^+$ $\;\underset{3}{\rightleftharpoons}\;$ 2 Glutamate + NAD(P)$^+$

Figure 3.21 The assimilation of ammonia into the amino acid glutamate. (a) (1) glutamate dehydrogenase (EC.1.4.1.3); (b) (2) glutamine synthetase (EC.6.3.1.2), (3) glutamate synthetase (EC.1.4.1.13)

It should be remembered that sulphur is also required by organisms for the formation of some amino acids, i.e., cysteine and methionine. Sulphate (SO_4^- is used to form cysteine and this requires a large amount of ATP. Details of the mechanism will not be given here.

3.9 Transport across the chloroplast membrane

It is assumed that the cell membranes (in this case for plant cells and photosynthetic bacteria) are permeable to a very large number of molecules. Generally membranes are only permeable to fairly small uncharged molecules, e.g., CO_2, unless there is a specific translocator (also referred to as transporter, carrier or permease) present in the membrane. Recently the transport across chloroplast membranes has been studied in some detail using methods similar to those used to study the permeability of the inner mitochondrial membrane.

Heldt and coworkers have studied transport across the chloroplast membrane and they define two membranes. The outer membrane is freely permeable to most molecules except large proteins and polysaccharides, e.g., enzymes and dextran; the inner membrane is selectively permeable and therefore acts as a barrier between the external space (the cytosol) and the stroma (including the thylakoid space).

3.9.1 Methods

The method most frequently used to study chloroplast permeability to anions is properly referred to as silicone layer filtering centrifugation, but may be called the 'spaces' technique. This method compares the extent of penetration of a radiolabelled anion into chloroplasts with that of tritiated (^{3}H) water (T_2O) and ^{14}C-labelled sorbitol (or sucrose). T_2O can penetrate throughout the whole chloroplast and sorbitol can only penetrate up to the inner membrane, i.e., not into the stroma (see figure 3.22).

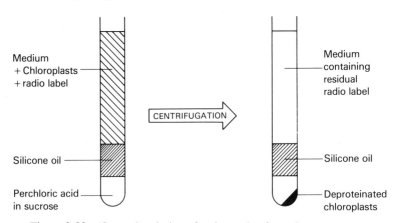

Figure 3.22 'Spaces' technique for the study of membrane transport

Small tubes (maximum capacity 400 μl) are prepared containing three layers of appropriate density such that the lower layer is perchloric acid, the middle layer is silicone oil and the uppermost layer is the incubation medium containing the chloroplasts and radiolabel (see figure 3.22). After the appropriate incubation time (usually up to 2 minutes) the transport can be terminated by rapid centrifugation. The chloroplasts pass through the silicone oil into the perchloric acid and from the radioactivity in the pellet (perchloric acid layer) it is possible to determine whether the labelled anion has penetrated into the chloroplast stroma. A better way of stopping the reaction is to use an inhibitor of the transporter, if one is known, and centrifugation can be carried out at a later time. This method is used for mitochondrial and whole cell permeability studies and is very convenient because only very small amounts of materials are needed.

3.9.2 The translocators

The translocators present in the chloroplast inner membrane are mainly anti-porters, i.e., they transport one ion in exchange for another of similar charge. There are some transporters which transfer one ion in exchange for another of dissimilar charge, e.g., adenine nucleotide and aspartate/glutamate transporters in mitochondria.

The *phosphate translocator* is specific for inorganic phosphate (Pi) and sugar phosphates with three carbons, i.e., 3 phosphoglyceric acid, dihydroxyacetone phosphate (DHAP) and glyceraldehyde 3 phosphate. Sugar phosphates with more than three carbons are not transported, neither are three carbon sugar phosphates where the phosphate is attached in the 2 position. The phosphates exchange 1 for 1 and each molecule transported inhibits the transport of the others. This translocator is inhibited by sulphydryl reagents as is the phosphate translocator in mitochondria, although the translocator in mitochondria is specific for Pi and hydroxyl ions. The *dicarboxylate translocator* transports malate, glutamate, aspartate, succinate and fumarate, and as with the phosphate translocator, each substrate for the translocator inhibits the transport of the other substrates. There is an *adenine nucleotide translocator* in the chloroplast membrane which exchanges ATP going into the chloroplast for ADP coming out of the chloroplast. This is the complete reverse of the exchange in mitochondria and the chloroplast transporter is much less active, presumably because a large amount of ATP produced in the chloroplasts is used in the stroma. It is not inhibited by atractyloside and it is thought to operate mainly in the dark to transfer ATP from glycolysis into the chloroplast for metabolism to continue there.

Glucose transport into the chloroplast is thought to proceed via a translocator similar to that found in erythrocyte membranes, i.e., it will not transport phosphates, has a K_m of 20 mM and is inhibited by dichloretin. Most amino acids, e.g., serine, glycine and phenylalanine probably diffuse across the chloroplast membrane. The rate of diffusion is slow, but the more hydrophobic the amino acid the faster it will cross the membrane.

There is still a great deal of research being carried out to characterize these translocators and to determine their role. Reducing equivalents which are produced in the chloroplast by photosynthetic electron transport need to be transferred into the cytosol for other biosynthetic reactions and so does the glucose produced by the Calvin cycle. The chloroplast membrane, like the mitochondrial membrane is impermeable to pyridine nucleotides (NAD(P)H)), and therefore the transfer of reducing equivalents into the cytosol has to be achieved by the use of shuttles (see figure 3.23). The oxaloacetate/malate shuttle can operate because there is a malate dehydrogenase in both compartments and oxaloacetate and malate can exchange on the dicarboxylate translocator. There are two malate dehydrogenase enzymes in the chloroplast, one NAD^+-linked and one $NADP^+$-linked, while that in the cytosol is NAD^+-linked. The $NADP^+$-linked enzyme is activated by light.

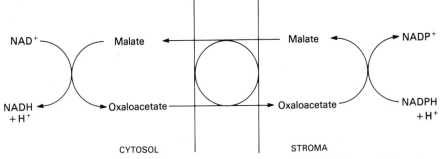

Figure 3.23   Oxaloacetate/malate shuttle for the transfer of reducing equivalents into the cytosol

Since aspartate, oxoglutarate, glutamate and malate can all be transferred across the membrane, and there is aspartate amino transferase in the stroma and the cytosol, reducing equivalents may also be transferred across the membrane via a shuttle analogous to the Borst cycle found in mitochondria.

Operating across the chloroplast membrane is another shuttle, referred to as the metabolite shuttle or the phosphoglycerate/DHAP shuttle (figure 3.24). This transfers NADPH from the stroma to the cytosol and also has the effect of transferring ATP as well. It is thought that the dihydroxyacetone phosphate (DHAP) exchanges for phosphoglycerate rather than glyceraldehyde 3

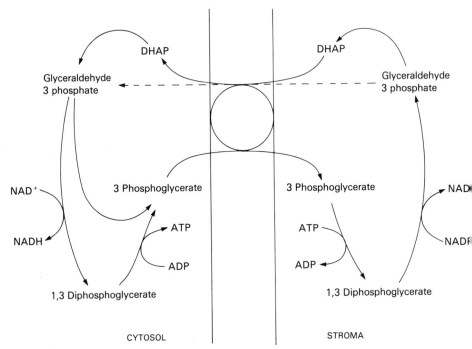

Figure 3.24   The phosphoglycerate/DHAP shuttle

phosphate, since the concentration of glyceraldehyde is low. The exchange is mediated by the phosphate translocator. This translocator also exchanges Pi for DHAP. In fact if Pi is added to a preparation of chloroplasts, then DHAP is released as well as some other triose phosphates. This may be important for the maintenance of the phosphate concentration in the chloroplast so that photophosphorylation and CO_2 fixation can continue. The DHAP is used for the biosynthesis of sucrose and starch in the cytosol as described in figure 3.11 (section 3.6).

3.10 Regulation of metabolism associated with photosynthetic systems

The rate of any enzymic reaction is affected by the concentration of substrates, the concentration of enzyme and the presence of hormones, activators and inhibitors. In simple terms the regulation of a metabolic pathway can be divided into short-term and long-term effects. Examples of long-term effects are the action of hormones, which affect a large number of cells for a long time, and the effects on protein synthesis, which, over a few days, may decrease or increase the concentration of enzymes of a particular metabolic pathway. Short-term effects occur in individual cells and are the responses of enzymes to rapid changes in the concentration of metabolites, coenzymes and ions. A very common controlling mechanism is feedback inhibition where an increase in the concentration of a product of the metabolic pathway inhibits an allosteric enzyme at the beginning of that pathway, e.g., ATP inhibits phosphofructokinase in glycolysis.

A very important factor in the regulation of photosynthetic systems will be light. The effect of light may not be direct but indirect, on the pH, the concentrations of Mg^{2+}, ATP and NAD(P)H and the reductive state of ferredoxin. The changes in these factors in the light and dark will be discussed below.

Photosynthetic systems have only a limited ability to alter the light intensity reaching them, e.g., by hormonal effects on leaf direction with respect to the sun. The effects of changing light intensity were discussed in chapter 1 (1.7). Generally, increases in light intensity give increased rates of photosynthesis until some other factor becomes rate limiting, e.g., CO_2.

Temperature changes affect all metabolic reactions. Photosynthetic systems do not have a sophisticated temperature regulating system like mammals and therefore a change in environmental temperature will effect rates of reactions and therefore growth. Increased temperatures will increase rates of reaction, but very high temperatures can be detrimental to enzymes. Increased temperature may also lead to closure of the stomata, to prevent water loss, which could limit the rate of CO_2 fixation because of CO_2 starvation. This problem has been solved by some tropical plants by the use of CAM (section 3.4). The overall effects of temperature and light on plant growth are discussed in chapter 1.

Table 3.1 shows some measurements of the pH of the stroma and thylakoid spaces in the light and in the dark. These measurements were made in spinach

Table 3.1   pH changes in the Chloroplast

	Light	Dark	pH change
Stroma pH	8.01	6.95	Increase of 1 unit
Thylakoid space pH	5.34	6.99	Decrease of 1.6 units

chloroplasts, but similar values were found in prokaryotic algae. These changes are partly due to proton transport into the thylakoid space in the light (section 2.9). There is also some light dependent H^+ transport from the stroma to the external medium, as there is a Mg^{2+}-dependent ATPase in the chloroplast envelope. The cytosol does not become as alkaline as the stroma (pH 8.0). The concentration of Mg^{2+} in the stroma increases in the light. Some of this Mg^{2+} movement is counter transport (for H^+) and compensates for 25 % of the charge difference across the thylakoid membrane created by the movement of H^+. There is no passive movement of Mg^{2+} across the chloroplast envelope. In the light the concentration of Mg^{2+} in pea plant chloroplasts reaches 10 mM and in the dark the concentration of free Mg^{2+} may be as low as 1 mM. There are variations in the values quoted in the literature for Mg^{2+} concentrations: some values are for total Mg^{2+}, i.e., bound and free, and some are just free Mg^{2+}. Illumination probably increases the stromal Mg^{2+}. Ca^{2+} is present in chloroplasts at concentrations similar to those of Mg^{2+}, but it is not free Ca^{2+}. The effects of ion changes on particular metabolic pathways in photosynthetic systems will be discussed shortly.

ATP and NAD(P)H are required for biosynthesis of carbohydrate (sections 3.2, 3.3 and 3.4) and light is essential for their production (see sections 2.8.2 and 2.10). ATP and reducing equivalents can be generated by oxidative metabolism in mitochondria in the dark; pH, Mg^{2+} and metabolite changes in the light may decrease this metabolism.

The concentration of adenine nucleotides (ATP, ADP and AMP), extracted with non-aqueous media, is approximately 2.5 mM in isolated plant chloroplasts. There is an active adenylate kinase present so changes in ATP concentration will lead to changes in the concentrations of ADP and AMP.

$$2\,ADP \rightleftharpoons ATP + AMP$$

On illumination of chloroplasts the concentration of ATP increases and that of ADP and AMP decreases. The concentration of inorganic phosphate (Pi) is ~10 mM and there are no large changes in its concentration. Although the phosphate potential of a system is generally calculated as [ATP]/([ADP][Pi]), for convenience here the [ATP]/[ADP] ratio will be measured. In the dark the ratio of ATP to ADP is 1.5 and this rises to 5 in the light in the chloroplast. In the cytosol the ratios are approximately 2.5 in the dark and 10 in the light, indicating a rapid transfer of ATP out of the chloroplast, probably by the action of the DHAP/phosphoglycerate shuttle (section 3.9.2). This increase in ATP will inhibit mitochondrial oxidative phosphorylation, since glycolysis will be inhibited (ATP action on phosphofructokinase) and increased ATP levels will inhibit citrate synthase and isocitrate dehydrogenase in the TCA cycle.

Concentrations of pyridine nucleotides present in chloroplasts are 0.5–1.5 mM, extracted by non-aqueous media. Some of this is in the reduced form in the dark: about 20 % of NADP was reduced and about 5 % of NAD was reduced. In the light these values change to 60–90 % reduced NADP and 30 % reduced NAD. The photosynthetic electron transport chain more readily reduces $NADP^+$ than NAD^+ which may account for the higher level of NADPH in the light. The NADH would probably be formed from the action of pyridine nucleotide-linked enzymes such as malate dehydrogenase. The concentration of reduced ferredoxin is affected by light, since it is reduced by electrons in the 'light reactions' (see figures 2.31, 2.32 and 2.33). Ferredoxin is a low molecular weight iron sulphur protein (MW $\sim$ 12 000) which can be reoxidized by $NAD(P)^+$ or nitrite (see chapter 2 and section 3.8).

The changes in some of the factors mentioned above in the light are shown in figure 3.25.

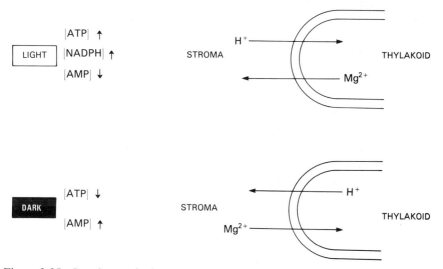

Figure 3.25   Ion changes in the stroma and thylakoid space induced by light; ↑ indicates increase in concentration and ↓ decrease in concentration

3.10.1 Regulation of the Calvin cycle

This cycle (see section 3.2) in the stroma of chloroplasts is regulated by several enzymes, but the main site of control is ribulose bisphosphate carboxylase (RBPC), the first reaction of the cycle. This is an allosteric enzyme containing, in chloroplasts, 8 catalytic and 8 regulatory subunits; the bacterial enzymes vary in their subunit content (section 3.7). The carboxylase activity of the enzyme is controlled by the concentration of its substrate CO_2, and it is inhibited by oxygen (see photorespiration, section 3.6).

The pH_{op} of the Calvin cycle is about 8.0 and there is little or no CO_2 fixation at pH 7.0. The pH changes in the stroma in the light (see table 3.1) will lead to an activation of CO_2 fixation. These pH changes and the changes in Mg^{2+} concentration, outlined above, activate several enzymes in the cycle. RBPC is activated by an increased affinity for its substrate (CO_2) in the presence of higher concentrations of Mg^{2+}. Phosphoglycerate kinase and fructose diphosphatase are also activated (see figure 3.26). The higher ATP concentration, lower AMP concentration and higher Mg^{2+} concentration (Mg^{2+} binds to adenine nucleotides) in the light, will stimulate the rates of enzymes requiring ATP, i.e., phosphoribulokinase and phosphoglycerate kinase. Glyceraldehyde phosphate dehydrogenase and phosphoribulose kinase may be activated in the light by the dissociation of oligmers to monomers.

The overall effect in the light is the stimulation of CO_2 fixation by RBPC and a stimulation of the regeneration of the ribulose bisphosphate and therefore the whole cycle is stimulated.

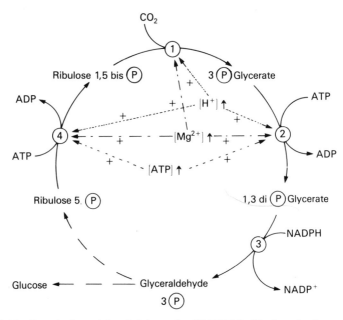

Figure 3.26   Regulation of the Calvin cycle: (1) RBPC, (2) phosphoglycerate kinase, (3) glyceraldehyde 3 phosphate dehydrogenase, (4) phosphoribulose kinase

The concentrations of Mg^{2+}, H^+ and adenine nucleotides in the dark inhibit the enzymes mentioned above. However, the cycle is left primed with ribulose bisphosphate, so that when the light intensity increases, CO_2 fixation can continue immediately.

When a plant ages there is a decrease in its ability to fix CO_2. This decrease parallels a decrease in the level of the enzyme ribulose bisphosphate carboxylase (RBPC), presumably because less enzyme is synthesized.

3.10.2 Regulation of C_4 CO_2 fixation and CAM

The controls described in the section above apply in plants exhibiting C_4- and crassulacean acid metabolism-type CO_2-fixation (sections 3.3 and 3.4), since the Calvin cycle operates in both cases. However, in the case of C_4 plants the first enzyme of the cycle is phosphoenol pyruvate (PEP) carboxylase, which is an allosteric enzyme controlled by feedback inhibition. It is a 4 subunit enzyme of total molecular weight 400 000, which uses bicarbonate (HCO_3^-) and not CO_2 as its substrate. The C_4 products of this pathway, oxaloacetate, malate and aspartate, inhibit PEP carboxylase. This inhibition is reduced as the pH and Mg^{2+} concentration increase, which occurs in the stroma in response to light. This enzyme is associated with the cytosol and the pH changes of the stroma are reflected there. The activity of the enzyme increases with increasing ATP/ADP ratio. This occurs in the light with increased photophosphorylation and movement of ATP into the cytosol via the phosphoglycerate/DHAP shuttle (section 3.9.2). This enzyme may also be controlled by some intermediates of the Calvin cycle, i.e., triose phosphates, phosphoglycerates and glucose 6 phosphate. An increase in the concentration of these intermediates would indicate that there was no longer a need for their synthesis for the biosynthesis of polysaccharides, amino acids and lipids, and therefore their inhibitory action on PEP carboxylase would inhibit the Calvin cycle.

In plants exhibiting CAM, malate acts as a feedback inhibitor of PEP carboxylase. The malate is compartmentalized in the vacuole at night, so that high concentrations of malate (100–200 mM) can be achieved without the carboxylase being inhibited. The affinity of NADP-linked malic enzyme (see figure 3.8) for malate decreases with a decrease in temperature, which occurs at night. Therefore malate would accumulate and inhibit the carboxylase at night unless it was transported into the vacuole.

Two other enzymes in C_4 plants that are inactivated in the dark and reactivated in the light are pyruvate phosphodikinase and NADP-linked malate dehydrogenase. The dikinase is inhibited by its products: PEP, AMP and pyrophosphate. It is thought to be located in the chloroplasts and this compartmentalization may control its activity.

The above two sections give a few main examples of control in photosynthetic CO_2 fixation. There are controls on the biosynthesis of polysaccharides, e.g., starch, amino acids and lipids, similar controls occur in other organisms too, but they will not be discussed here.

One metabolite that occurs in several of the metabolic pathways discussed is phosphoenol pyruvate (PEP). A summary of its central position in these metabolic pathways is shown in figure 3.27. When one metabolite has several metabolic fates, the enzymes involved in its metabolism are usually quite carefully controlled. This applies to the enzyme involved in PEP metabolism.

Light has an important effect on C_3 and C_4 CO_2 fixation. In high light intensities more ribulose bisphosphate carboxylase and PEP carboxylase are synthesized.

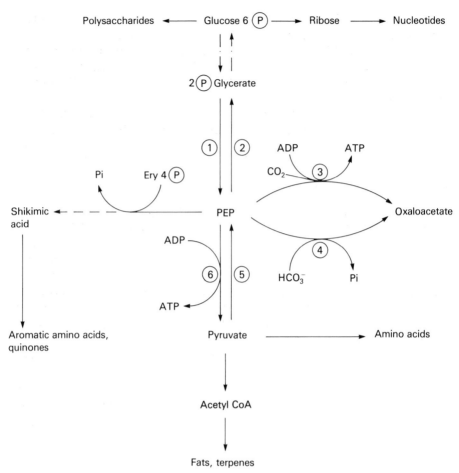

Figure 3.27 Central role of PEP in metabolism in photosynthetic systems: (1) and (2) enolase (EC.4.2.1.11), (3) PEP carboxykinase (EC.4.1.1.49), (4) PEP carboxylase, (5) pyruvate phosphodikinase, (6) pyruvate kinase (EC.2.7.1.40), (7) first step in the formation of shikimic acid, important for aromatic amino acid biosynthesis

Photorespiration is inhibited by increased CO_2 concentrations; light stimulates it and photorespiration increases linearly with oxygen concentration.

3.10.3 Regulation of nitrogen metabolism

Like CO_2 fixation, nitrogen fixation requires a large amount of energy (ATP). All the steps involved in converting atmospheric nitrogen (N_2) to the amino group of an amino acid (section 3.8) require energy and therefore the controls on nitrogen metabolism act to prevent wasteful use of ATP. A high ADP/ATP ratio will inhibit nitrogenase and nitrate reductase activity. Ammonia is an important controlling factor, and it will accumulate if there is no oxoglutarate or glutamate available to react with it. Increased NH_3 concentrations repress the biosynthesis of nitrogenase and nitrate and nitrite reductases.

In the light the ADP/ATP ratio will decrease and nitrogenase will be activated, as will other reactions requiring ATP. There will also be increased levels of oxoglutarate, due to active CO_2 fixation, and therefore NH_3 concentrations will decrease.

The control of amino acid biosynthesis is too involved to discuss in detail. An important regulating mechanism in amino acid metabolism is feedback inhibition.

3.10.4 The direct effects of light

So far in this section the indirect effect of light on photosynthetic metabolic pathways has been discussed, i.e., the changes of pH, Mg^{2+}, ATP, NAD(P)H and reduced ferredoxin concentrations. However, light has been shown to have a direct effect on some photosynthetic enzymes. The effect is the generation of $-SH$ groups. Two enzymes which are affected this way are fructose diphosphatase and NADP-linked malate dehydrogenase. This reduction of disulphide bridges by light may account for the dissociation of oligomers to monomers in glyceraldehyde phosphate dehydrogenase and phosphoribulose kinase. This reduction is probably mediated by thioredoxin, which transfers electrons from reduced ferredoxin (reduced by photosynthetic electron transport) to the enzyme.

There is some debate about the role of this reduction in photosynthetic systems. It may ensure that accessible sulphhydryl groups are kept reduced in the presence of oxidants generated by photosynthetic reactions, e.g., hydrogen peroxide (H_2O_2).

3.11 Biosynthesis of chlorophyll, gibberellic acid and phytochrome

The scope of this book does not allow us to discuss all aspects of the metabolism which occurs in photosynthetic cells. In this section the biosynthesis of three important groups of molecules involved in photosynthesis is discussed. These are chlorophyll, which is involved in the trapping of light, the plant hormone gibberellic acid and phytochrome which is an important mediator of plant responses.

3.11.1 Chlorophyll

The structure of chlorophylls was discussed in section 2.2.1 (figure 2.1). The constant structure in all chlorophylls consists of 4 pyrrole rings linked by methene bridges and referred to as a porphin. When various groups are attached it becomes a porphyrin, the most common being protoporphyrin IX (figure 3.28). This forms the basis of haem, the prosthetic group for haemoproteins, e.g., haemoglobin, vitamin B_{12} and cytochromes as well as chlorophylls.

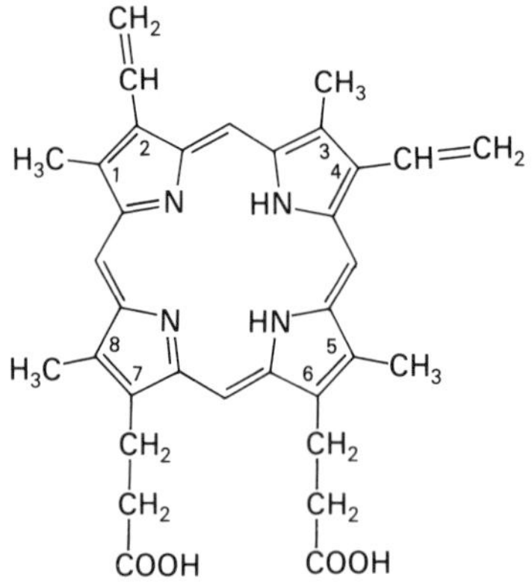

Figure 3.28   Structure of protoporphyrin IX

The biosynthesis of protoporphyrin IX is almost identical in plants, animal tissues and bacteria and is shown briefly in figure 3.29. 5-aminolevulinic acid (ALA) is formed from succinyl CoA and glycine by the enzyme ALA synthetase. ALA dehydratase then dehydrogenates 2 molecules of ALA to give porphobilinogen (PBG) containing a pyrrole group. Four molecules of PBG are then used to make the porphyrin, urogen III. Three further enzyme catalysed steps give rise to protoporphyrin IX. ALA synthetase has been found in photosynthetic bacteria but there is very little evidence for its activity in plants. The enzyme for succinyl CoA synthesis (oxoglutarate dehydrogenase) resides in plant mitochondria and it has been suggested that ALA was synthesized in the mitochondria and transported into chloroplasts. However, this is unlikely as the chloroplast membrane is impermeable to ALA. An alternative route for the synthesis of ALA in chloroplasts has been suggested (figure 3.29), where ALA is synthesized from glutamate and from oxoglutarate. There are a large number of enzyme catalysed steps, which convert protoporphyrin IX to chlorophyll a and b, etc. Some of these steps are discussed in section 4.2.4 and several of them, as would be expected, are controlled by light. ALA formation stops in the dark. Plants become etiolated in the dark and when they are returned to the light ALA accumulates and this inhibits the hydratase. In the dark etioplasts are present containing prolamellar bodies. Only in the light do these become chloroplasts and as the thylakoids form, protochlorophyllide a is converted to chlorophyll a. This means that although ALA is not being synthesized in the dark, some more immediate precursors of the chlorophylls are present. The biosynthesis of the chlorophylls occurs simultaneously with membrane biosynthesis.

3.11.2 Gibberellic acid

There are several groups of hormones associated with photosynthetic systems, which must therefore be synthesized in these systems. Here only 1 group will be

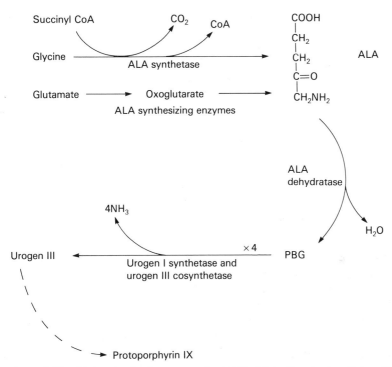

Figure 3.29   Biosynthesis of protoporphyrin IX. ALA: 5-aminolevulinic acid

examined. Gibberellins are plant hormones with structures similar to that of gibberellic acid (figure 3.30), which is usually abbreviated to GA$_3$. They were first discovered in about 1926 as the metabolic product of the fungus *Gillberella funjikuroi*, which caused infected rice plants to become very elongated. A great many gibberellins have been isolated and all have a similar effect to varying extents, i.e., increase in plant growth, stem length increases but the number of internodes does not alter. Recent studies indicate that many of these gibberellins could be intermediates of biosynthesis or breakdown products.

Gibberellins occur in large amounts mainly in the growing points and in young expanding leaves, and it has been suggested that these are the sites of biosynthesis. The starting point for biosynthesis of gibberellic acid is mevalonic acid, MVA (figure 3.31), which is also an important precursor for terpenoids, e.g., carotenoids, and sterols, as it is a source of 5 carbon isoprene units. Mevalonic acid is formed from 3 molecules of acetyl CoA in the cytosol. The

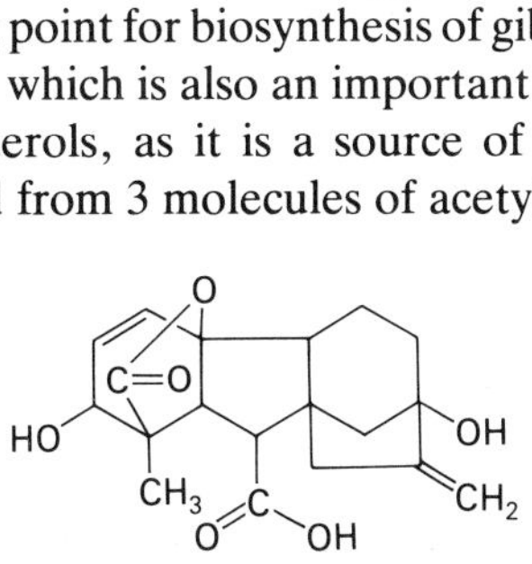

Figure 3.30   Structure of gibberellic acid (GA$_3$)

details of the biosynthesis of gibberellins from mevalonic acid is well characterized in plants and microorganisms, thanks mainly to the use of radio-labelled substrates in cell free systems.

There is some evidence for the compartmentalization of gibberellins in plastids, which may be important in the regulation of their concentration. It has also been shown that phytochrome (section 3.11.3) regulates their biosynthesis and release from plastids.

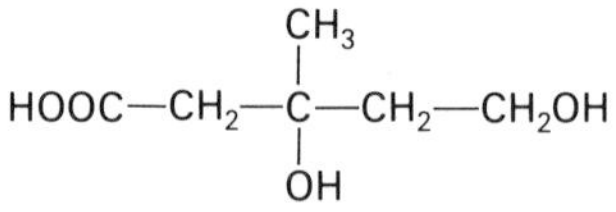

$$HOOC-CH_2-\overset{\overset{\displaystyle CH_3}{\displaystyle |}}{\underset{\underset{\displaystyle OH}{\displaystyle |}}{C}}-CH_2-CH_2OH$$

Figure 3.31   Structure of mevalonic acid (MVA)

3.11.3 Phytochrome

Phytochrome is a protein with a chromophore attached which mediates a large number of changes in plants. The chromophore is an open chain tetrapyrrole which can exist in two forms, Pr and Pfr. Pr is blue, inactive and absorbs red (660 nm) light while Pfr is green, active and absorbs far-red (730 nm) light. The highest concentrations of phytochrome are found in roots and shoots (meristematic tissues).

Some of the effects possibly mediated by phytochrome are the change in orientation of chloroplasts in high and low light intensities, the biosynthesis and release of gibberellins and the control of enzymes such as ribulose bisphosphate carboxylase, nitrate reductase and peroxidase. In its active, Pfr, form it can bind to subcellular structures and it can be found in many membrane preparations. This has led to the suggestion that phytochrome's primary site of action is membranes.

There is some phytochrome present in dry seeds and during germination *de novo* synthesis of the Pr form occurs. The bio-synthesis of the tetrapyrrole group is probably by similar metabolic pathways to that described for proto-porphyrin IX (section 3.11.1).

Suggested further reading

Metabolism

Davies, D. D. (1979). The central role of PEP in plant metabolism. *Ann. Rev. Plant Physiol.*, **30**, 131

Fuller, R. C. (1978). Photosynthetic carbon metabolism in green and purple bacteria, in Clayton, R. K. and Sistrom, W. R. (Eds.), *The Photosynthetic Bacteria*, Plenum Press, New York, p. 691

Hatch, M. D. and Slack, C. R. (1970). The C_4 pathway of photosynthesis. *Ann. Rev. Plant Physiol.*, **21**, 141

Lorimer, G. H. (1981). The carboxylation and oxygenation of ribulose 1,5 bisphosphate, *Ann. Rev. Plant Physiol.*, **32**, 349

Osmond, C. B. (1978). Crassulacean acid metabolism (CAM). *Ann. Rev. Plant Physiol.*, **29**, 379

Woolhouse, H. W. (1978). Light-gathering and carbon assimilation processes in Photosynthesis; their adaptive modifications and significance for agriculture, *Endeavour*, **2**, 35

Zelitch, I. (1977). Pathways of carbon fixation in green plants. *Ann. Rev. Biochem.*, **46**, 123

Photorespiration

Goldsworthy, A. (1976). *Photorespiration*, Carolina Biology Reader no. 80 (Ed. J. J. Head), Burlington, N. Carolina

Rao, K. K. and Hall, D. O. (1982). Photorespiration, *J. Biol. Educ.*, **16**, 167

Nitrogen metabolism

Postgate, J. (1978). *Nitrogen Fixation*, Studies in Biology, no. 92, Edward Arnold, London

Quebedeaux, B. (1979). Symbiotic N_2 fixation and its relationship to photosynthetic carbon fixation in higher plants, in M. Gibbs and E. Latzko (Eds.), *Encyclopaedia of Plant Physiology* New Series, vol. **6**, Photosynthesis II p. 472

Transport

Heber, U. (1974). Metabolite exchange between chloroplasts and cytoplasm. *Ann. Rev. Plant Physiol.*, **25**, 393

Heldt, H. W. (1976). Metabolite exchange in intact spinach chloroplasts, in J. Barber (Ed.), *The Intact Chloroplast*, Elsevier/North Holland Biomedical Press, Amsterdam, chap. 6

Regulation

Hatch, M. D. (1978). Regulation of enzymes in C_4 photosynthesis, *Curr. Topics in Cell Reg.*, **14**, 1

Kelly, G. J., Latzko, E. and Gibbs, M. (1976). Regulatory aspects of photosynthetic carbon metabolism, *Ann. Rev. Plant Physiol.*, **27**, 181

Krause, G. H. and Heber, U. (1976). Energetics of intact chloroplasts, in J. Barber (Ed.), *The Intact Chloroplast*, Elsevier/North Holland Biomedical Press, Amsterdam, chap. 5

Mortenson, L. E. (1978). Regulation of nitrogen fixation. *Curr. Topics in Cell Reg.*, **13**, 179

Smith, F. A. and Raven, J. A. (1979). Intracellular pH and its regulation. *Ann. Rev. Plant Physiol.*, **30**, 289

Walker, D. A. (1976). Regulatory mechanisms in photosynthetic carbon metabolism. *Curr. Topics in Cell Reg.*, **11**, 203

Relevant to section 3.11

Harel, E. (1978). Chlorophyll biosynthesis and its control. *Prog. Phytochemistry*, **5**, 127

Heddon, P., MacMillan, J. and Phinney, B. O. (1978). Metabolism of gibberellins, *Ann. Rev. Plant Physiol.*, **29**, 149

Marne, D. (1977). Phytochrome. Membranes as possible site of primary action. *Ann. Rev. Plant Physiol.*, **28**, 173

4

The Origin and Assembly of Chloroplasts

4.1 Introduction

We have seen in previous chapters that chloroplasts are comparatively large organelles with a complex ultrastructure and functional organization. The problem of how these organelles originate and develop is one which has stimulated speculation for almost one hundred years and one in which there has been considerable intensification of investigation in the last fifteen years or so. The questions that have been asked are essentially the following four:

(1) Do developing chloroplasts arise *de novo* or do they derive from some undifferentiated precursor organelle?
(2) Is there continuity of either the postulated precursor organelles or the mature chloroplasts in terms of growth and division of the organelles?
(3) How is the development, structure and functioning of chloroplasts controlled at the genetic level?
(4) What biochemical activities are involved in processing of the genetic information

A distinct, but related problem has been the evolutionary origin of these organelles.

This chapter devotes itself to an analysis of the present state of understanding in these areas. Understanding of some of these problems, in particular those concerned with genetic control and expression, is advancing extremely rapidly at the present time as the problems become amenable to the new and powerful techniques which have previously been applied to the genetic analysis of prokaryotes. Nevertheless, there are still many unresolved problems.

4.2 Growth and development of chloroplasts

4.2.1 Cellular origin of chloroplasts

As photosynthetic cells grow and divide it is clearly necessary that the numbers and total mass of the chloroplasts must also be increased to keep pace. There are two fundamental possibilities for their origin within the cell. Either the chloroplasts or their immature precursors are assembled *de novo* within the cell or they arise only by growth and division of other chloroplasts or chloroplast precursor structures. The latter view is now accepted virtually universally. It was first stated in fairly clear terms by Schimper (1885) although other workers had previously observed what we now know to be plastids in the process of division forty years previously. Schimper stated that: 'With regard to their reproduction and chemistry they behave much more like independent organisms than part of the cytoplasmic body.'

Other early studies have provided clear evidence of the division of chloroplasts in a variety of plant species. The clearest demonstrations have come from studies with a number of algae. In the unicellular flagellate alga *Chromulina*, which contains a single chloroplast and mitochondrion, each of these undergoes a doubling in volume between cell divisions and divides at cell division to provide a single representative of both organelles in each daughter cell. Particularly convincing have been the cinematographic demonstrations of chloroplast division in the alga *Nitella*. However, division of mature chloroplasts cannot be the only way in which new chloroplasts arise. The meristematic regions of virtually all plants from algae to angiosperms are composed of colourless cells devoid of chloroplasts and yet the cells of these regions will give rise to a variety of cell types some of which are rich in mature chloroplasts. Clearly, in these cases, the mature chloroplasts must be elaborated from simpler structures. An idea that persisted for several years was that chloroplasts arose from mitochondria. In retrospect it is clear that this suggestion arose because of the difficulty in distinguishing in the light microscope between mitochondria and the immature plastids which are now referred to as proplastids. Even with electron microscopy it is sometimes difficult to distinguish mitochondria and proplastids unless serial sections are examined.

Another view that was put forward firstly at the beginning of this century was that plastids arose from the nucleus. This view was given added weight by the observations of Bell in the 1960s who used the electronmicroscope to study the process of oogenesis in the fern *Pteridium aquilinum*. It was observed that during maturation of the egg cell the mitochondria and plastids became vacuolated and swollen. This was thought to indicate degeneration of the organelles. During the same period of development the egg cell nucleus also begins to evaginate, forming vesicles which separate from the nucleus. It was suggested that these vesicles developed into new plastids and mitochondria and that at this stage a completely new generation of organelles is assembled *de novo*. It was also considered that this was a general phenomenon which would occur also in plants other than ferns. These conclusions have received little support from the work of others. Firstly, the general view has been that there is no degeneration of organelles although, possibly, there may be a measure of dedifferentiation. There is little support also for the view that the vesicles evaginating from the nucleus developed into plastids or mitochondria. These are now thought to remain as distinct vesicles at the periphery of the egg or to break down releasing their nucleoplasmic contents into the cytoplasm.

In summary, there is no convincing evidence that developing plastids arise from anything other than proplastids. It is now generally accepted that chloroplast continuity is maintained by growth and division of the immature proplastids which under appropriate conditions will undergo differentiation to mature chloroplasts.

The physical continuity of plastids is also strongly supported by plant breeding studies. Classical studies on variegated plants, in which the colourless areas had defective chloroplasts, clearly demonstrated a non-chromosomal pattern of inheritance of the defect (section 4.5.1). This implied an extra-nuclear genetic determinant passed from generation to generation. It was natural to assume that this determinant, whose effect seemed to be on the chloroplast, should reside in the chloroplast itself.

4.2.2 Mechanics and control of chloroplast division

In most cases division of plastids seems to occur by the formation of a constriction in the middle of the plastid which is finally completely severed. This is presumably brought about by some active contractile system. Some plastids seem to undergo unequal divisions in which a relatively undifferentiated vesicle may bud off from a plastid, the vesicle having the potential to develop at a later stage into a fully functional plastid.

A variety of factors have been shown experimentally to influence the rate of plastid division, although, in some cases, it is not clear how important these controls are under normal circumstances. Light, of course, is exceptionally important in chloroplast development and it does appear to play a role, if somewhat less dramatic, in division. Proplastids do not seem to require light for their replication as they will multiply under conditions of zero illumination. With mature chloroplasts, however, the general rule seems to be that light intensity and duration of illumination do appear to influence the numbers and mass of chloroplasts, although there seems to be a saturation level of illumination. Many studies have been done with cultured spinach leaf discs which have been shown not to require light for division of chloroplasts. These plant cultures can be supported in the dark with sucrose as a carbon source and, under these conditions, a tripling in the number of chloroplasts occurs within nine days. There appears to be no increase in chloroplast mass, however, as the size of the chloroplasts decreases considerably. Under diurnal illumination (14 hour day) a similar increase in chloroplast numbers and decrease in size occurs during the dark periods with an increase in both size and numbers during light periods. High light intensities of blue, red or green light stimulate chloroplast replication in light-grown spinach leaf discs, but there are no observable effects of low intensity red-light as might be expected if the replication was under the control of the phytochrome system. In contrast, dark-grown broad bean plastids show a stimulation of plastid division by low-intensity red light which can be reversed by far-red light.

Temperature and nutrition also affect plastid replication. This is slowed down at lower temperatures (10–15°C) when there is an increase in chloroplast size. Replication is also slowed down under conditions of iron or manganese deficiencies.

4.2.3 Plastid development

The chloroplasts of the mature green areas of the plant originate from proplastids. The process of development to the mature chloroplast has undergone considerable investigation from both biochemical and morphological standpoints. The study of chloroplast maturation under normal conditions of diurnal illumination present a number of difficulties. Firstly, the chloroplasts will present a considerable degree of variation of structure throughout the plant and even the leaf. This leads to difficulty in understanding the precise developmental sequence at the morphological level and makes studies on the biochemical sequence of chloroplast development almost impossible. Also the periodic light/dark transitions make it very difficult to isolate the role of light in the developmental sequence. As light is an exceptionally important factor in development, this is a considerable drawback. The alternative approach, which emphasizes the important role of light, has been to grow seedlings for a period of time in the dark and then to transfer them to continuous light. Growth of seedlings in the dark results in the plastids developing all to the same stage. These plastids, arrested in their development because of lack of light are referred to as etioplasts. On subsequent illumination the etioplasts all begin to develop synchronously and ultimately become typical chloroplasts. This synchronous development permits unambiguous determination of a developmental sequence and facilitates biochemical investigation. However, this rather artificial experimental approach to the study of chloroplast development raises questions concerning whether observations made in this system can be applied to the developmental processes occurring under 'normal' conditions of illumination. The process of development of proplastids to chloroplasts under the continuous dark to continuous light regime can be seen as a two-phase process – the conversion of proplastids to etioplasts followed by the conversion of etioplasts to chloroplasts.

The conversion of proplastids to etioplasts presents a similar problem of heterogeneity and asynchrony of development as does chloroplast development under diurnal conditions. However, the process has been studied in some detail by Bradbeer. The proplastid is an organelle with an outer limiting double membrane and having a diameter something of the order of $1\,\mu m$ (figure 4.1). In general there is little or no internal organization although sparse internal membrane sheet structures can be visualized, sometimes appearing as invaginations of the inner membrane. During dark growth the amount of membrane structure within the proplastid increases enormously leading to formation of typical etioplasts (figure 4.2). The most notable feature of etioplasts is the presence within the organelle of a geometrically organized network of tubules referred to as the prolamellar body. Continuous with the tubules of the prolamellar body are several membrane sheets referred to as either porous lamel-

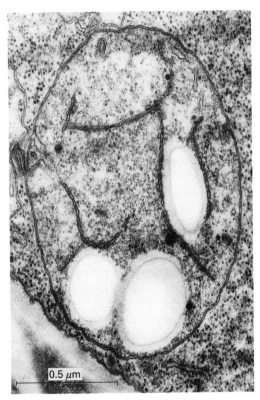

Figure 4.1 Electron micrograph of proplastid from dark-grown *Phaseolus vulgaris*. (Reproduced from Moules and Bradbeer, *Proceedings of 3rd International Congress of Photosynthesis*, 1973, Vol. III, 1867–1876 by permission of Elsevier Biomedical Press)

lar sheets or prothylakoids. The prolamellar body has a regular repeating structure. It has been described in terms of hexagonally arranged tubules, each sheet of hexagons being joined to neighbouring sheets by other tubules arising from the angles of the hexagons – three alternate nodes on each hexagon connecting with a hexagon in the sheet above, the other three with the sheet below. A diagrammatic representation of a small part of this array is shown in figure 4.3. Why these structures form this paracrystalline arrangement is not at all clear although it has been shown that solubilized prolamellar bodies will reaggregate back to highly branched tubules. Bradbeer showed that in etioplast development in dark-grown broad bean (*Phaseolus vulgaris*) the first membranes to appear within the proplastid appear to be of the porous lamellar sheet type and suggests that the prolamellar body derives from a condensation of these sheets of membrane. Membrane synthesis seems to cease after fourteen days or so in this system.

Transformation of etioplasts to chloroplasts can be studied with much greater precision because of the synchronous development occurring following the onset of illumination. After thirty minutes or so of illumination the prolamella

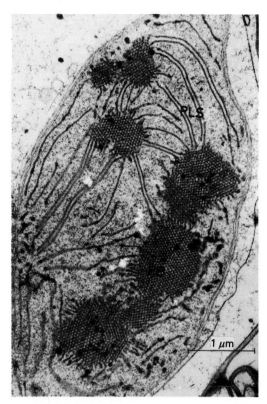

Figure 4.2 Electron micrograph of etioplast in dark-grown *Phaseolus vulgaris*. PB: prolamellar body. PLS: porous lamellar sheet. (Reproduced from J. W. Bradbeer *et al.*, *New Phytologist*, 1974, **73**, 263–270 by permission of the publishers)

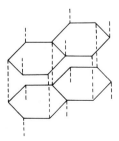

Figure 4.3 Diagrammatic representation of tubular structure of small part of a prolamellar body. Solid line hexagonal sheets are perpendicular to the plane of the paper. Broken lines linking sheets are parallel to the plane of the paper

body starts to break down from its apparently organized to a more disorganized arrangement, as seen by electron microscopy following osmium tetroxide/glutaraldehyde fixation. A more rapid disorganization seen with permanganate fixation (within one minute of the onset of illumination) has been attributed to a fixation artefact; nevertheless, it is still presumed to reflect a fundamental change in the structural stability of the prolamellar body. Between thirty minutes and four hours the prolamellar body completely disappears. There seems to be a rough approximation between the surface area of the prolamellar body and that of the prothylakoids present at four hours, indicating a precursor/product relationship between these two structures. Thereafter, there is a considerable increase in the thylakoid membrane area which must be the result of new membrane synthesis. Adpression of the thylakoids becomes apparent at approximately ten hours and formation of grana begins. These events occur at an early stage in membrane synthesis.

The nature of the illumination is important in determining the precise structural organization of the chloroplasts achieved. Chloroplasts from plants grown under low light intensity tend to be larger with more thylakoids and larger grana than those in plants grown in conditions of high illumination. When the illumination regime is millisecond flashes at fifteen minute intervals, the chloroplasts which develop are green but lack grana, having parallel but unfused thylakoids. The main light wavelength effects which have been observed indicate a role for phytochrome in chloroplast development. Activated phytochrome promotes plastid growth and formation of plastid membrane as well as the effects on plastid division that we have already remarked on (section 4.2.2). These effects are reversed by far-red light. There also seem to be a number of other photoreceptors controlling plastid development.

The changes occurring in development of chloroplasts in higher plants is paralleled to a limiting extent in unicellular algae. Most studies have been performed using *Chlamydomonas reinhardtii*. As we have already noted division of chloroplasts in unicellular algae appears to occur under normal conditions of illumination exclusively at the mature chloroplast stage. However, *Chlamydomonas* can be induced to grow and divide in the dark on a medium with acetate as a carbon source. Following transition from an autotrophic to a heterotrophic mode of existence the organisms undergo a process of chloroplast dedifferentiation referred to as degreening. During degreening it has been shown that the internal membranes of the chloroplasts do not undergo active destruction as degreening does not occur to a significant extent in the absence of cell division. When cells are dividing, however, there is an exponential dilution of chloroplast material leading to cells virtually devoid of chlorophyll but it has been shown that all chloroplast components are not diluted at a similar rate, as total membrane content decreases more slowly than chlorophyll level. This implies active synthesis of chloroplast membranes – a situation which is clearly essential if the cells are to retain the ability to regenerate functional chloroplasts on subsequent illumination. The synthesis of membranes and components normally associated with chloroplast membranes in dark-grown cells is supported by biochemical investigations (section 4.2.4). Following degreening a structure superficially similar to a prolamellar body can be found in some *Chlamydomonas* (figure 4.4). However, closer inspection shows that although this appears to be a complex network of intermeshed tubules it lacks the paracrystalline organization seen in prolamellar bodies of higher plants. Furthermore, on illumination, it lacks the rapid changes in structure and stability seen in a prolamellar body following illumination of etiolated higher plants and it fails to undergo a complete conversion to prothylakoids even after prolonged illumination. A similar type of disorganized prolamellar-body-like structure has also been seen in dark-grown *Euglena gracilis*.

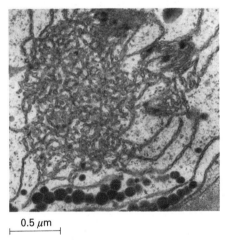

0.5 μm

Figure 4.4   Electron micrograph of prolamellar-like body in degreening *Chlamydomonas reinhardtii*. (Reproduced from Friedberg *et al.*, *The Journal of Cell Biology*, 1971, **50**, 268–275 by copyright permission of The Rockefeller University Press)

Following illumination of degreened algae, greening occurs. This can take place in the absence of cell division and results in the conversion of β-carotene-coloured yellow-orange individuals to typical green coloured ones. In *Euglena*, at least, it has been suggested that some membranes formed during degreening originate from the ground substance (stroma) of the etioplast. It seems unlikely that this is correct and it is possible that examination of serial sections would indicate an origin of thylakoid membranes either by invagination of the inner membrane of the etioplast envelope or by conversion of the prolamellar-body-like structure. In *Euglena*, thylakoids begin to stack and form grana approximately four hours following illumination and this process continues throughout the whole subsequent greening process. At fourteen hours the shape of the plastid begins to change from a relatively spherical to an elongate structure.

4.2.4 Biochemical parameters of chloroplast development

From the biochemical viewpoint etioplasts are complex structures containing the majority of the components found in mature chloroplasts although in many cases in reduced quantities. The major biochemical differences between chloroplasts and etioplasts are the absences in the latter of chlorophylls *a* and *b* and chlorophyll–protein complexes and the absence of cytochrome $b559_{HP}$.

During greening the photosynthetic enzymes already present undergo increases in quantity and activity. The two subunits of the ribulose bisphosphate carboxylase complex (E.C.4.1.1.39) start to increase in quantity immediately following illumination. This was found by gel electrophoretic investigation. The activity of the enzyme complex does not start to increase, however, until a significant lag has occurred. The mechanism of activation of the complex is thought to involve a small molecule (molecular weight $\simeq 5000$). Other enzymes known to undergo activation during greening are phosphoribulose kinase (E.C.2.7.1.19) and triose phosphate dehydrogenase (E.C.1.2.1.13). These enzymes can be activated *in vitro* by ATP, NADPH or sulphydryl reagents. As with ribulose bisphosphate carboxylase, the activity of the enzymes does not increase before a lag of several hours following the onset of greening. If the enzyme extracts are preincubated with ATP, however, enzyme activity is seen to increase from the onset of illumination. In the case of phosphoribulose kinase, during the first five days of greening there is a ninety-fold increase in total enzyme activity. This is composed of a ten-fold increase in enzyme protein and a nine-fold increase in specific activity.

The photosynthetic components absent from etioplasts have a light-requiring step in their synthesis. One such light-dependent step is the conversion of protochlorophyllide to chlorophyllide (figure 4.5). Protochlorophyllide is present in the membranes of the prolamellar body as a protein complex. It can be visualized as particles equivalent in size to that of the extracted complex

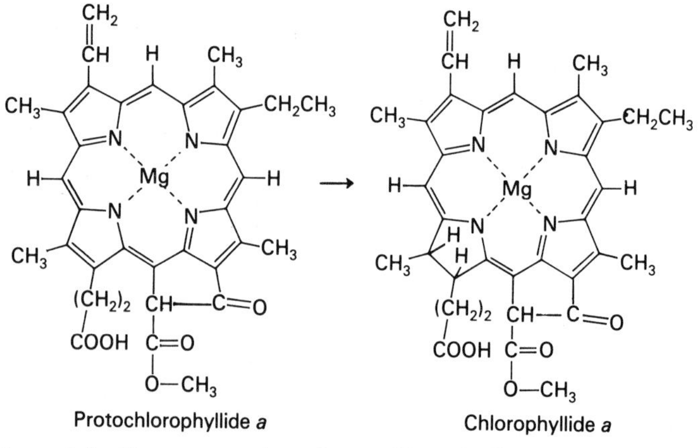

Protochlorophyllide *a*   Chlorophyllide *a*

Figure 4.5   Photoconversion of protochlorophyllide to chlorophyllide

by electron microscopy of negatively-stained sections or by freeze-fracture electron microscopy. Three spectroscopic forms of protochlorophyllide can be distinguished with absorption maxima at 628, 637 and 650 nm, respectively, indicating different aggregate forms. The forms with 637 and 650 nm maxima undergo photoreduction to chlorophyllide on illumination. Two hydrogens are added stereospecifically to positions 7 and 8 on the porphyrin ring (figure 4.5). The reaction requires the protochlorophyllide to be bound to its protein holochrome. This reaction occurs extremely rapidly (within a fraction of a second of the onset of illumination in 6-day-old, dark-grown barley seedlings). The protein moiety is thought to act as an enzyme (NADPH-protochlorophyllide oxidoreductase) and to be involved in the following reaction sequence:

$$\text{NADPH} + \text{Protochlorophyllide} + \text{Enzyme} \rightleftharpoons [\text{Protochlorophyllide-enzyme-NADPH}]$$

$$[\text{Protochlorophyllide-enzyme-NADPH}] \overset{\text{light}}{\rightleftharpoons} [\text{Chlorophyllide-enzyme-NADP}^+]$$

$$[\text{Chlorophyllide-enzyme-NADP}^+] \rightleftharpoons \text{Chlorophyllide} + \text{Enzyme} + \text{NADP}^+$$

The conversion of chlorophyllide to chlorophyll a involves esterification of a phytyl residue (from phytyl alcohol) to the side chain on carbon atom 7 (see figure 4.5). This reaction is not light dependent and occurs relatively slowly, resulting in a lag before appearance of significant quantities of chlorophyll following illumination. The enzyme involved is chlorophyllase (E.C.3.1.1.4). Chlorophyll b is formed from chlorophyll a in the aggregates. Immediately following illumination only chlorophyll a can be detected but the ratio of chlorophyll a/chlorophyll b falls rapidly as the chlorophyll a to b conversion occurs until the level approaches that present in mature chloroplasts.

Photosynthetic oxygen evolution, absent from etioplasts, becomes detectable at thirty minutes following illumination. This increases dramatically during the next couple of hours, after which the total level of photosynthetic oxygen production on a plant fresh-weight basis is comparable with that of mature plants. On the basis of chlorophyll content the specific activity at this stage in development is very much higher than that in mature plants. Both photosystems become active at an early stage although they are not as large at this early stage as they become later on. It seems that the photoconversion apparatus is assembled at an early point but that the light-harvesting units are not completed until some time later. Initially (after 15 minutes or so), photosystem 1 activity is higher than photosystem 2 activity. This is presumably because the electron transport elements associated with photosystem 1 (cytochromes f, b_6 and $b559_{LP}$, plastocyanin and ferredoxin) are all present in etioplasts. This might permit early cyclic photophosphorylation with the ATP generated being made available for biosynthesis of chloroplast components and activation of the various ATP-activated photosynthetic enzymes. The light dependence of cytochrome $b559_{HP}$ synthesis could possibly limit photosystem 2 activity although it has not been possible to find a correlation between photosynthetic oxygen evolution and cytochrome $b559_{HP}$ levels, suggesting that some other component may be limiting at this stage.

Lipid composition during greening does not change markedly in the first five hours following illumination but subsequently there is a significant increase in galactolipids. This coincides with the onset of grana formation.

The biochemical changes during degreening in *Chlamydomonas* have also been studied. The most dramatic occurrence is the immediate cessation of chlorophyll synthesis. P700 levels appear to decline more rapidly than chlorophyll levels. Photosystem 2 actively falls off at a similar rate to chlorophyll dilution whereas there is actually an increase in the specific activity of photosystem 1 (relative to chlorophyll content). Many of the other components fall in content although to a smaller extent than chlorophyll. This suggests that these components continue to be synthesized in the dark albeit at a lower level than in light. These enzyme changes are accompanied by structural changes as indicated by the fall in P700 relative to chlorophyll *a* and other spectroscopic changes and also in the susceptibility of various membrane proteins to digestion by trypsin.

4.3 Chloroplast DNA

4.3.1 Discovery and properties of chloroplast DNA

One of the most exciting discoveries in the field of chloroplast biology was the discovery within chloroplasts of DNA physically and chemically distinguishable from that of the nucleus. This discovery added a completely new dimension to the problem of chloroplast growth and development, presenting the possibility of an integrated control of the functioning of an organelle by two distinct genetic systems.

The first discovery of chloroplast DNA (ctDNA) came in cytological studies using the DNA-specific Feulgen stain. Feulgen staining bodies were found in chloroplasts which were sensitive to enzyme degradation by DNAse. Subsequently the presence of DNA in isolated chloroplasts was demonstrated chemically. The most convincing demonstration came from analytical ultracentrifugation of whole-cell and chloroplast DNA from *Chlamydomonas*. This showed a DNA component with a density 1.695 g cm^{-3} which comprised 6 % of whole-cell DNA and was considerably enriched in isolated chloroplasts. Figure 4.6 shows the data from such an experiment. DNA has subsequently been found in chloroplasts of all plants where it has been looked for.

The ctDNAs from a wide variety of plant species show a remarkable uniformity of kinetic complexity. Kinetic complexity refers to the total quantity of sequence-distinct DNA and is a measure of the potential genetic information in the system under study. The usual approach to determination of kinetic complexity is to bring about heat denaturation of extracted DNA and to follow the rate of reannealing during cooling, generally by spectrophotometry. The kinetics of reassociation can be used to determine kinetic complexity. In all cases

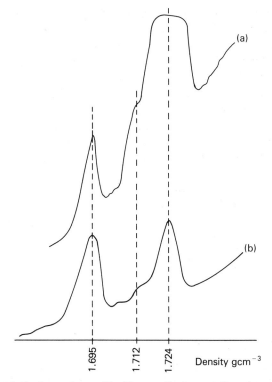

Figure 4.6 Analytical caesium chloride gradient centrifugation of DNA from *Chlamydomonas reinhardtii*. (a) Whole cell DNA. (b) DNA from partially-purified chloroplasts. Peaks are as follows: 1.695 g cm³ chloroplast DNA; 1.712 g cm⁻³ mitochondrial DNA; 1.724 nuclear DNA

(except giant unicellular algae of the genus *Acetabularia*) the kinetic complexity is fairly close to 1×10^8 molecular weight. Some workers have described what appears to be, from its rapid rate of renaturation, a highly reiterated segment of chloroplast DNA with a molecular weight in the order of 10^6 to 10^7. Not all workers have observed this rapid-renaturing fraction of ctDNA and it has been suggested that the rapid renaturation may result from incomplete denaturation during heating, although more recent work on the mapping of ctDNA from *Zea mays* and *Chlamydomonas reinhardtii* by restriction enzyme analysis suggest that there may be some reiterated sequences within the main sequence in these.

As previously mentioned, members of the genus *Acetabularia* appear to have ctDNAs of kinetic complexity some 10–15 times those observed in other plants. The reason for this is not altogether clear. Two suggestions are that the longer ctDNA may also be present in very small quantities in other plants or that *Acetabularia* chloroplasts have a larger degree of autonomy than those of other species.

Chloroplast DNA appears to exist mainly as circular molecules (figure 4.7).

130

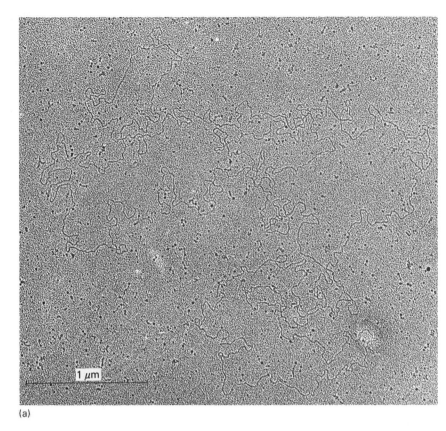

(a)

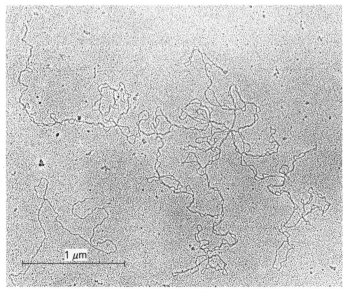

(b)

This has been demonstrated for a number of species of higher plants (beans, lettuce, maize, oats, peas and spinach) and also for the algae *Chlamydomonas reinhardtii* and *Euglena gracilis* and the liverwort *Sphaerocarpus castellanii*. Despite the difficulties inherent in preparation for electron microscopy of ctDNA molecules without shearing, careful techniques have led to the production of preparations with as much as 90 % of the ctDNA in a circular form. Electron microscopy of such preparations in the presence of a DNA standard of suitable known size allows estimation of the length of individual chloroplast DNA molecules. By these techniques it has been shown that the ctDNAs of individual plant species are homogeneous in size although differing in sizes from species to species. Reported molecular contour measurements vary in the main from about 38 up to 75 μm, from species to species (table 4.1). Linear

Table 4.1   Densities and contours of chloroplast DNA

	Contour (μm)	Density (gcm^{-3})
Chlamydomonas reinhardtii	62	1.695
Euglena gracilis	40	1.685
Lactuca sativa	43	1.697
Nicotiana tabacum	44	1.697
Pisum sativa	39	1.697
Spinacea oleracea	46	1.696

molecules of the same order of size can also be found and these are thought to arise from a double-strand break in the circular DNA. Shorter linear molecules probably arise by fragmentation while the occasional linear or circular molecules exceeding this size may arise by oligomer formation. The contour lengths of circular ctDNAs from higher plants usually fall in a narrower band between 38 and 46 μm, while the size of *Acetabularia* chloroplast DNA falls well outside the general range (38–75 μm) and seems to be close to 200 μm. *Acetabularia* ctDNA may also be linear. Although the precise relationship between molecular weight and contour lengths varies slightly with experimental conditions (thus the necessity to include a DNA standard of known contour in a preparation), in general, 1 μm length of DNA approximately corresponds to 1.9×10^6 molecular weight. Consequently, molecular weights of ctDNAs from higher plants vary from about 9.0×10^7 up to 10.6×10^7. These figures agree remarkably with the figures for kinetic complexity of ctDNAs which leads to the conclusion that all ctDNA molecules in a single cell have the same base sequences. This conclusion has been reinforced by partial denaturation

Figure 4.7   Electron micrographs of rotary-shadowed chloroplast DNA molecules. (a) Circular molecule from *Euglena gracilis* (40.5 μm contour), (b) supercoiled circular molecule from *Spinacia oleracea* (42 μm contour). (Reproduced from J. E. Manning *et al.*, *PNAS*, 1971, **68**, 1170 and J. E. Manning *et al. The Journal of Cell Biology*, 1972, **53**, 594–601 by copyright permission of Dr D. R. Wolstenholme and The Rockefeller University Press)

mapping of pea ctDNA where all molecules exhibit the same denaturation pattern and consequently probably have similar base sequences. This technique involves treating DNA for a short time with alkali in the presence of formaldehyde. As the denaturation treatment is relatively mild only the most susceptible regions of the DNA (usually A-T rich areas) denature. The presence of formaldehyde prevents renaturation and on electron microscopic examination A-T rich regions on the ctDNA show up as separated single strands. Restriction enzyme analysis also supports the conclusion that ctDNA molecules are a homogeneous population, as treatment of ctDNAs with restriction enzymes leads to fragmentation of the DNA into a collection of pieces whose total length is approximately the same as the contour of the circular molecule. This has been done for a number of plant species.

Chloroplasts appear to be polyploid with respect to ctDNA, the mean number of copies of each DNA circle being about 20 to 100 per chloroplast. This has been surmized from studies on the total DNA content of chloroplasts. Similarly, electron microscopy and autoradiography have provided evidence for the presence of several DNA-containing areas in chloroplasts. There seem to be usually less than 20 DNA-containing regions per chloroplast, suggesting that each region may contain a number of ctDNA molecules. These regions are referred to as nucleoids.

The buoyant densities of ctDNAs from higher plants are fairly uniform, varying between 1.694 and 1.698 g cm^{-3} (table 4.1), indicating a GC content of about 36–40 %. This uniformity of base composition is not found in ctDNAs of algae where densities ranging from 1.685 g cm^{-3} (*Euglena gracilis*) up to 1.696 g cm^{-3} (*Acetabularia mediterranea*) have been observed. The uniformity in base composition of ctDNA in higher plants should not be considered as evidence for similarities in base sequences. Restriction fragments of ctDNAs from plants of different genera or families do not show common fragmentation patterns but similar fragmentation patterns can be observed between ctDNAs of species within a genus. The buoyant density of ctDNA in any higher plant is normally very close to the density of its nuclear DNA and the significance of this fact is not clear.

The ctDNAs of some plants have been shown to contain relatively large inverted repeat sequences separated by non-repeated segments. In those higher plants in which this structure has been found the repeated segment is usually of the order of 25 kilobase pairs in length. In *Chlamydomonas* it is 19 kilobase pairs. The structure is not always present in ctDNA – it is absent in *Pisum sativa* and *Euglena gracilis* – so if it has an important function it is not a universal one.

There are two other features of interest in the chemistry of ctDNA. Firstly ctDNAs appear to contain a number of ribonucleotides within their sequence. This has been shown by their responses to ribonuclease treatment and also their alkali-lability. Kinetic studies on ctDNA breakdown in the presence of alkali have indicated the presence of 18 ± 2 ribonucleotides per molecule in pea and spinach ctDNAs and 12 ± 2 in lettuce ctDNA. Electron microscopy of the alkali-produced fragments of pea ctDNA indicated that the molecule breaks at 19 unique sites; the implication is that each unique site contains just a single ribonucleotide. To date, the significance of these sites is not clear.

Another feature of ctDNAs is their apparent absence of methylation in most cases in higher plants and algae. There has been a report of a very low level of methylation in a portion of ctDNA in spinach but this is thought to be due to nuclear contamination.

4.3.2 Replication of chloroplast DNA

It is clear that the DNA found within chloroplasts originates from replication of other ctDNA molecules and is not delivered to the chloroplast from some external source. It has been shown by autoradiography that chloroplasts are capable of incorporating [³H]-thymidine into discrete areas within the chloroplast, presumably the nucleoids (figure 4.8). DNA polymerase has also been shown to be present in isolated tobacco or spinach chloroplasts and the product of polymerase activity has been shown to be complementary in base sequence to ctDNA.

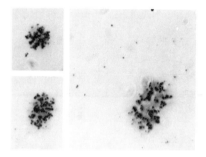

Figure 4.8 Autoradiographs of *Beta vulgaris* chloroplasts isolated from [³H]-thymidine-treated leaves. (Reproduced from R. Herrmann, *Planta*, 1970, **90**, 80–96 by permission of Springer-Verlag, Heidelberg)

Chiang and Sueoka used ¹⁵N/¹⁴N density transfer experiments of the Meselson and Stahl kind to demonstrate that replication of chloroplast DNA is semi-conservative in *Chlamydomonas reinhardtii*. They also showed that there was a distinct period in the cell cycle during which chloroplast DNA synthesis occurred and that this was different from the period when nuclear DNA was being synthesized. Other workers showed that *Euglena gracilis* chloroplast DNA was also synthesized at one particular period in the cell cycle but that, in this case, the synthetic period coincided with that for nuclear DNA. This type of experiment has not been done for higher plants because of the difficulties inherent in obtaining a synchronous population of cells and the similarities in densities of their nuclear and chloroplast DNAs.

The mechanics of the DNA replication process have been studied in higher plants and algae by electron microscopy. Both Cairns 'θ'-form and rolling-

134

circle 'σ'-form replication intermediates have been observed. In maize and pea ctDNA replication seems to initiate by formation of a pair of single-strand displacement loops (D-loops), one in each strand of the DNA (figure 4.9). The two D-loops extend towards each other and on meeting fuse to form one loop of a Cairns-type replication intermediate. Replication then proceeds bi-directionally until the daughter ctDNA molecules are complete. Sigma (σ)-form replication intermediates may be generated from the daughter molecules if the newly synthesized strand is not immediately ligated. The possibility that the σ-forms may in reality be broken Cairns-type molecules is unlikely because there appears to be a unique initiation site and the tail of the sigma may exceed in length the contour of a single circle-features characteristic for a rolling-circle model for DNA replication. It is thought that the two mechanisms for replication may represent responses to different cellular demands for chloroplast DNA synthesis. Rapid chloroplast replication may utilize the prolific 'printing-press' approach of the rolling circle, but for normal division-rate ctDNA replication may be controlled via the Cairns mechanism, which would require separate initiation events for each molecule replicated.

There is certainly a considerable measure of modulation applied to ctDNA replication. Under circumstances where chloroplast growth and division rate is sufficiently fast to increase chloroplast numbers in cells in higher plants, the

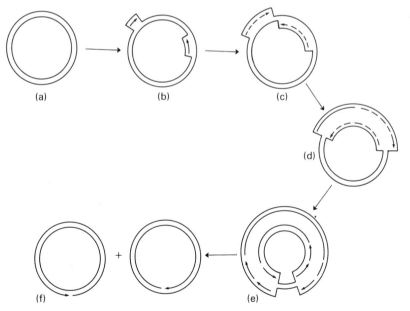

Figure 4.9   Postulated mechanism of replication of chloroplast DNA. (a) Close-circular duplex parent molecule. (b) Molecule containing 2 displacement loops (D-loops). (c) Molecule containing expanded-D-loops. (d) and (e) Cairns-type replicative intermediates. (f) Progeny molecules with newly synthesized strands still to be ligated.
(Adapted from Kolodner and Tewari, (*Nature*, **256**, 708, 1975) with permission)

rate of ctDNA synthesis can exceed the rate of nuclear DNA synthesis. It has been suggested that an integrated control of ctDNA replication is applied based on rates of chloroplast division and nuclear DNA replication.

4.4 Expression of the chloroplast genome

4.4.1 Transcription of chloroplast DNA

The transcription of any genetic information encoded in chloroplast DNA into RNA sequences requires the operation of a DNA-dependent RNA polymerase system. This enzyme has been isolated from chloroplasts from a number of plants (*Chlamydomonas*, maize, tobacco and wheat) and the enzymic characteristics of these enzymes have been investigated. In *Chlamydomonas* the polymerase has been shown to comprise five kinds of subunit of molecular weights: 96 000, 93 000, 52 000, 51 000 and 39 000. Clearly, the degree of complexity is equivalent to that in prokaryotic or nuclear RNA polymerase systems in this case but the subunit sizes do not correspond to those of chloroplast RNA polymerases from other sources. Maize chloroplast DNA-dependent RNA polymerase appears to have two main subunits, molecular weights 180 000 and 140 000, plus several smaller subunits. This is rather similar to yeast mitochondrial RNA polymerase.

RNA polymerases from chloroplasts vary in their responses to the action of the prokaryotic inhibitor rifampicin. Most seem to have some sensitivity to rifampicin or to related compounds, but experimental conditions seem to be important. They are not sensitive to α-amanitin which inhibits some nuclear polymerases. Control of polymerase activity may resemble prokaryotic systems as the requirement of a sigma-factor for promoter recognition by the polymerase has been reported.

4.4.2 Chloroplast protein-synthesizing equipment

As well as containing DNA and a transcription system for that DNA, chloroplasts contain a complete outfit for synthesizing proteins within the organelle. In the main, the elements of this system can be readily distinguished from their counterparts in the cytoplasm. The most fundamental component of the chloroplast protein synthesizing system is the chloroplast ribosome. In contrast to the typical 80S ribosomes found in eukaryotic cytoplasm, chloroplast ribosomes have a sedimentation value of 70S. This is similar to ribosomes of prokaryotes. Chloroplast ribosomes share a number of other properties with bacterial ribosomes. Both are sensitive in their protein-synthesizing activities to chloramphenicol and a variety of other inhibitors which block bacterial protein synthesis. These include erythromycin, carbomycin, oleandomycin and lincomycin which, like chloramphenicol, affect the large ribosomal subunit, and kanamycin, neamine and spectinomycin which affect the small ribosomal subunit. In contrast, chloroplast ribosomes are insensitive to cytoplasmic protein synthesis inhibitors such as cycloheximide and emetine. Like prokaryotic ribosomes, chloroplast ribosomes dissociate into large and small subunits at low Mg^{2+} concentration and reassociate on restoration of Mg^{2+} levels. Eukaryotic cytoplasmic ribosomes, on the other hand, require prolonged dialysis in the absence of Mg^{2+} to bring about dissociation and do not reassociate on replacement of Mg^{2+}. 70S chloroplast ribosomes dissociate into large and small subunits (50S and 35S in tobacco; 50S and 30S in *Euglena*). The sedimentation values of these subunits is similar to those from prokaryotic ribosomes. The similarities between chloroplast and prokaryotic ribosomes is perhaps most strikingly illustrated by the fact that hybrid ribosomes with protein synthetic activity can be prepared. Small subunits of *Euglena* chloroplast ribosomes direct phenylalanine uptake into trichloroacetic acid precipitatable material in the presence of poly U when combined with *Escherichia coli* large subunits and supernatant factors. The reciprocal combination of subunits is not active. With spinach chloroplast ribosomes either type of hybrid ribosome is active in protein synthesis. This functional hybrid ribosome formation does not occur using subunits of chloroplast ribosomes with those of plant cytoplasmic ribosomes.

Ribosomes are composed largely of RNA and a variety of proteins. Chloroplast ribosomes have been shown to contain four RNA species: 23S, 5S and 4.5S RNAs in the large subunits and 16S RNA in the small subunit. (This contrasts with 25S, 5.8S and 5S from the large subunit and 18S from the small subunit of plant cytoplasmic ribosomes.) The proteins of chloroplast ribosomes also display a prokaryotic pattern; there are about 55 in all, 34 in the large subunit and 21 in the small subunit.

Chloroplasts contain the necessary tRNAs and aminoacyl-tRNA synthetase enzymes necessary for protein synthesis. These are distinct from their cytoplasmic counterparts (and also from their counterparts within the mitochondria). Two-dimensional gel electrophoresis of spinach chloroplast tRNAs has fractionated these into 35 components. 27 to tRNAs corresponding to 16 amino acids have been identified: alanine, asparagine, histidine, lysine, phenylalanine, proline, tryptophan and tyrosine each have a single isoacceptor tRNA; two isoacceptors are found for arginine, glycine, isoleucine, threonine and valine. Leucine, methionine and valine have three isoacceptors. Eight other tRNA spots have yet to be assigned to amino acids. It seems likely then that chloroplasts contain a full set of tRNAs. This also appears to be the case in maize chloroplasts. The base sequences of chloroplast tRNAs seem to have closer similarity to prokaryotic tRNAs than the eukaryotic cytoplasmic ones.

With respect to aminoacyl-tRNA synthetases only a few have been demonstrated in isolated chloroplasts of higher plants or *Euglena* which has led to the suggestion that tRNAs may become aminoacylated under the influence of enzymes imported from the cytoplasm. There is some experimental evidence in favour of this possibility in *Euglena*.

The presence of messenger RNAs in chloroplasts is also established. RNA extracted from spinach chloroplasts has been shown to direct the synthesis of discrete polypeptides in a cell-free protein synthesizing system from *Escherichia coli*. The bulk of chloroplast-located mRNAs in spinach seems to be lacking in polyA tails (or having only very short ones). A similar situation has been reported in *Euglena* although it is suggested that a proportion of polyadenylated RNA from maize is located in the chloroplast. There is some doubt about the ability of cytoplasmic ribosomes to translate chloroplast mRNA and chloroplast ribosomes to translate mRNA of nuclear origin. It has been reported that spinach chloroplast mRNA can be translated, *in vitro*, using an *E. coli* (70S) ribosomal system but not using a wheat-germ (80S) ribosomal system. RNA from chloroplasts of the aquatic angiosperm *Spirodela oligorhiza* does appear to be translated in a wheat-germ system. *In vivo* it has generally been considered that only chloroplast DNA transcripts are translated on chloroplast ribosomes, although there is a recent suggestion that as much as 70 % of protein synthesis in isolated spinach chloroplasts may be directed by transcripts of nuclear origin.

Initiation of chloroplast protein synthesis involves N-formyl methionyl tRNA, this being similar to the situation in prokaryotes but contrasting with the involvement of a non-formylated methionyl tRNA in cytoplasmic ribosomes of plant cells. Elongation factors are necessary for the binding of aminoacyl tRNAs to the ribosomes and the translocation step of protein synthesis. Once again these factors are very similar to those in prokaryotes. Little is known about the termination process of protein synthesis in chloroplasts.

4.4.3 *Coding and synthesis of chloroplast components*

As indicated in the previous sections the question of whether nuclear transcripts can be translated on chloroplast ribosomes or chloroplast transcripts on cytoplasmic ribosomes is still not absolutely established. The chloroplast proteins whose coding and synthetic origin are established are either encoded in nuclear DNA and synthesized on cytoplasmic 80S ribosomes, or encoded in ctDNA and translated on chloroplast ribosomes. Many workers believe that this pattern is an absolute rule. Chloroplasts contain several hundred different proteins or polypeptides. The coding potential of chloroplast DNA suggests that the genetic information within the molecule could make a significant contribution to specification of chloroplast structure and function. A molecular weight of DNA of 10^8 is approximately equivalent to 150 kilobase pairs which could be expected to specify several hundred average size polypeptide molecules. Not all of the chloroplast genome specifies protein structure however, since a proportion of the genome encodes the sequences of ribosomal and transfer RNAs. The chloroplast DNAs from a variety of sources (broad bean, lettuce, maize, pea, spinach and wheat) contain two copies each of the DNA sequences complementary to 23S and 16S RNAs. This has been shown by DNA/RNA hybridization. *Euglena gracilis* contains three copies of these genes. Genes for 5S and 4.5S RNAs have also been ascribed to chloroplast

DNAs, probably also more than one copy per genome. There have also been some reports suggesting that extra copies of chloroplast rRNA genes may reside in the nucleus. The close linkage of the different chloroplast rRNA genes (see section 4.5) and the discovery of large precursor RNAs suggests that all of the RNAs may be initially transcribed as a single precursor RNA.

RNA/DNA hybridization has also been used to show that a number of transfer RNA genes are located on chloroplast DNA. Twenty-six cistrons for tRNAs have been found in *Euglena gracilis* ctDNA. Similar numbers of tRNA genes have been found in ctDNAs from a number of higher plant species. It seems likely that there may be more such genes – perhaps one for each tRNA species found in chloroplasts – but technical reasons may have prevented their recognition. There has been a number of approaches to the study of chloroplast coding and synthesis of the organelle proteins. One of these makes use of antibiotics with more or less specific effects on cytoplasmic and chloroplast ribosomal protein synthesis. Cycloheximide is generally used as an inhibitor of cytoplasmic protein synthesis; this drug is without effect on chloroplast protein synthesis. Chloramphenicol is usually used to block chloroplast protein synthesis. Interpretation of results obtained using these antibiotics must be made with caution. In the first place the drug may not be absolutely specific in its action. It may also bring about inhibition of protein synthesis in one compartment as an indirect effect of inhibition of synthesis of some particular protein in another compartment. Nevertheless, these studies have provided much useful information which complements findings using alternative techniques. In particular, synthesis of certain proteins in the presence of chloramphenicol is good evidence that they are *not* synthesized on chloroplast ribosomes. This approach has shown that almost all of the soluble proteins of the chloroplast are synthesized on cytoplasmic ribosomes. This includes most of the enzymes of the Calvin cycle, aminoacyl-tRNA synthetases, RNA and DNA polymerases. The only soluble protein positively identified as being synthesized on chloroplast ribosomes is the large subunit of ribulose bisphosphate carboxylase-oxygenase (E.C.4.1.1.39 – fraction 1 protein). This enzyme complex has a dual function of carboxylation of ribulose bisphosphate (in photosynthetic carbon fixation, see section 3.2.1) or oxidation of ribulose bisphosphate to 3-phosphoglyceric acid and 2-phosphoglycolic acid (in photorespiration, see section 3.5). Ribulose bisphosphate carboxylase can represent up to 50 % of the total soluble protein in leaf extracts and is thought to be possibly the most abundant protein in nature. Chloramphenicol has been shown to block specifically the synthesis of the large subunit of this protein whereas cycloheximide preferentially blocks the synthesis of the small subunit. Such a dual synthetic origin of subunits of particular enzyme complexes has also been observed in the cytochrome oxidase and oligomycin-sensitive ATPase complexes from mitochondria, each containing subunits manufactured on cytoplasmic and mitochondrial ribosomes.

The compound 2-(4-methyl-2,6-dinitroanilino)-N-methylpropionamide (MDMP), an inhibitor of cytoplasmic protein synthesis on 80S ribosomes, has been found to block the synthesis of both large and small subunits of ribulose bisphosphate carboxylase in intact pea leaves. It is thought that this may be because the small subunit (synthesized on cytoplasmic ribosomes) acts as an initiation factor for the transcription or translation of the large subunit. This would seem to be an effective means of coordinating the rates of synthesis of the two subunits.

In addition to the large subunit of ribulose bisphosphate carboxylase, inhibitor studies have given indication that other proteins may be synthesized on chloroplast ribosomes. These are membrane components, including the photosystem 1 chlorophyll protein and some chloroplast ribosomal proteins.

A second approach to the problem of which proteins are synthesized on chloroplast ribosomes has been the study of the proteins synthesized in isolated chloroplasts. Light is used to drive the photosynthetic ATP generating system as an energy source for protein synthesis. This approach does not have the disadvantages previously mentioned for the antibiotic approach. Here the assumption is made that proteins synthesized in isolated chloroplasts must be coded by ctDNA and synthesized on chloroplast ribosomes. A possible source of error here may be the contribution of extra-chloroplast ribosomes not removed by the preparative procedure. However, this should be fairly small since using light as an energy source only intact organelles will synthesize ATP and this will only be available for internal protein synthesis. Another possible problem is that cytoplasmic factors may be essential for the synthesis in the chloroplast of certain proteins. Incubation of isolated chloroplasts with radioactively-labelled amino acids, in the presence of light, followed by polyacrylamide gel electrophoresis, confirmed inhibitor study indications that the large subunit of ribulose bisphosphate carboxylase is made on chloroplast ribosomes. The identity of the product as the ribulose bisphosphate carboxylase large subunit was confirmed by analysis of the peptide products of tryptic digestion. A number of other peaks of protein synthetic activity have been observed including the photosystem 1 chlorophyll protein (molecular weight 110 000) in broad bean (but not in pea) and a 32 000–35 000 molecular weight membrane protein synthesized in isolated maize and pea chloroplasts. The synthesis of photosystem 2 chlorophyll protein was also claimed in isolated broad bean leaf chloroplasts but as this was found to be sensitive to cycloheximide it was attributed either to contaminating 80S cytoplasmic ribosomes or possibly to a specific class of 80S ribosomes involved particularly in the synthesis of this protein. Other proteins claimed to be synthesized in isolated chloroplasts include three of the five subunits of chloroplast CF_1 ATPase, cytochromes b_{559} and f and two elongation factors for protein synthesis, although the evidence for these is not so cut and dried as for the proteins previously mentioned.

A similar approach to this one has been the use of chloroplast mRNA to direct the synthesis of proteins in a heterologous protein synthesizing system. Chloroplast mRNAs from a number of sources have been shown to direct the synthesis of the large subunit of ribulose bisphosphate carboxylase and the 32 000 molecular weight membrane protein – both of which were shown by tryptic peptide analysis to be similar to their counterparts synthesized in isolated chloroplasts.

These types of study provide some supporting evidence for the coding origin of these proteins if it is assumed that freshly synthesized mRNA is used in chloroplast translation processes. However, it is difficult to rule out completely the possibility that some proteins may be specified in isolated chloroplasts by long-lived messengers of nuclear origin. The problem of chloroplast coding has been tackled by more direct approaches. The first of these uses ctDNA as the information source for a linked transcription–translation system. Proteins synthesized in such a system would be encoded in the DNA template provided. Either total chloroplast DNA or cloned fragments of ctDNA have been employed. Use of a linked *E. coli* RNA polymerase/rabbit reticulocyte protein synthesizing system has shown the presence in ctDNA of a gene specifying the sequence of ribulose bisphosphate carboxylase large subunit.

A combined biochemical-genetic approach has also provided clear cut evidence for the location of the large subunit ribulose bisphosphate carboxylase gene on ctDNA. This has been done using a number of *Nicotiana* species. It has been demonstrated that ctDNA is transmitted maternally in *Nicotiana* crosses, i.e., the ctDNA of *Nicotiana* species is provided only by the gamete from the female parent, with no contribution from the male gamete. The location of the gene for the large subunit of ribulose bisphosphate carboxylase on ctDNA has been conclusively demonstrated in this system. The large subunit shows an extra tryptic peptide in *N. gossei* over the equivalent subunit from *N. tabacum*. When *N. gossei* is employed as the female parent in an interspecific cross, F_1 plants have ribulose bisphosphate carboxylase large subunit containing the extra peptide, i.e., the protein structure is similar to that in *N. gossei*. When *N. gossei* is the male parent and *N. tabacum* the female parent the tryptic digest of the large subunit protein resembles that from *N. tabacum*. Similar experiments desgined to determine the coding origin of the small subunit and employing species *N. tabacum* and *N. glutinosa*, which have contrasting tryptic digestion patterns in this protein subunit, do not show such maternal inheritance and so the gene for this protein is considered to be located in nuclear DNA. Similar conclusions have been derived from studies where subunit polypeptides were characterized by isoelectric focusing and amino acid sequencing studies.

Two possibly related problems which have yet to be answered are the mechanisms of uptake of cytoplasmically synthesized polypeptides into the stroma and chloroplastically synthesized (or cytoplasmically synthesized) polypeptides into the thylakoid membranes. It has been suggested that a signal mechanism operates which is similar to that proposed for transport of proteins from ribosomes to the cisternae of the endoplasmic reticulum. The signal hypothesis proposes a sequence of bases in the mRNA following the AUG initiation codon, and specifying a signal sequence of amino acids at the N-terminal end of the protein to be transported across the membrane. The signal sequence is thought to be recognized by a membrane receptor protein which guides the nascent protein – signal sequence first – through some kind of tunnel in the membrane. If this type of system operates in chloroplasts one would expect to see ribosomal associations with the various chloroplast membranes. To date, there is no clear evidence for association of cytoplasmic ribosomes with the chloroplast envelope. However, there have been reports that when *Chlamydomonas* is treated with chloramphenicol to prevent dissociation of mRNA/ribosome/protein/membrane complexes, ribosomes can be found to be bound to thylakoid membranes. These can be removed from the membranes by puromycin and high salt concentrations which is considered to be evidence that the ribosomes are engaged in protein synthesis. The quantity of bound ribosomes responds to changes in the illumination regime in a manner which suggests an involvement in membrane assembly.

4.5 Chloroplast genetics

4.5.1 Extra-chromosomal inheritance

The fact that certain genetic traits appeared to be specified by genes which did not obey the usual Mendelian rules of inheritance has been known virtually since the rediscovery of Mendelian principles early in this century. There are two major indicators of non-chromosomal inheritance of traits. One indication is when non-Mendelian ratios of progeny characteristics are observed in crosses; a second is when there are significant differences in the results of reciprocal crosses. The early discoveries illustrated both of these features. Correns and Baur published simultaneously the results of their independent work on the inheritance of leaf variegation. Variegation results from the segmentation of plants into areas which have normal plastids and are green and those which have undeveloped plastids and are either yellow or white. Correns worked with *Mirabilis jalopa* and found a strictly maternal inheritance of the plastid defect (figure 4.10). When a flower on a white (or yellow) region of the plant is pollinated the resulting seeds give rise only to white seedlings irrespective of the nature of the pollen parent. Similarly, if a flower on a green branch is pollinated the resultant progeny will be all green. Some individual branches show variegation, however, and if a flower on such a branch is pollinated the progeny may be green, white or variegated. In all cases the nature of the pollen parent does not affect the result. In other words the gene whose mutation has brought about the plastid defect is transmitted only through the maternal line. A similar situation has already been described in members of the *Nicotiana* genus (section 4.4.2) and seems to be a fairly widespread phenomenon. However, there are also a large number of plant species where such maternal or uniparental transmission of non-nuclear genes for plastid development does not occur and where the pollen makes a significant contribution, i.e., biparental inheritance occurs. Baur worked with *Pelargonium*, for example, and found that both parents contributed to variegation in the progeny but that these were not produced in strictly Mendelian ratios and ratios were different in reciprocal crosses (figure 4.11). Subsequent work in this area has shown an exceedingly

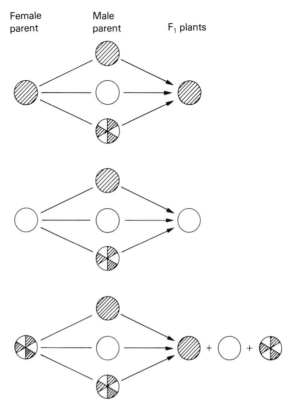

Figure 4.10 Inheritance of defective plastids in *Mirabilis jalopa*. Shaded, open and sectored circles represent green, white or variegated plants (or regions of plants) respectively

complex pattern of interactions between nuclear and non-nuclear genes in chloroplast development.

Although it seems logical that the extra-chromosomal determinants, which are affected in plants with variegated leaves, should be located on chloroplast DNA, it has been exceptionally difficult to prove this rigorously. The problem is to eliminate the possible contributions of either mitochondrial genes or genes associated with some still undiscovered cytoplasmic genomic structure. Transformation of white cells to green by transplantation of chloroplasts into protoplasts of white cells has been achieved but there is still the possibility that some contaminating genetic structure might be responsible. Absolute demonstration of the involvement of ctDNA would require identification of the primary lesion and the defective gene product and demonstration of chloroplast DNA coding of that protein.

A mutant type found in a number of higher plants is male sterility. In a wide range of cases this male sterility has been shown to be non-nuclear in the mode of inheritance. This situation has usually been attributed to mutations in

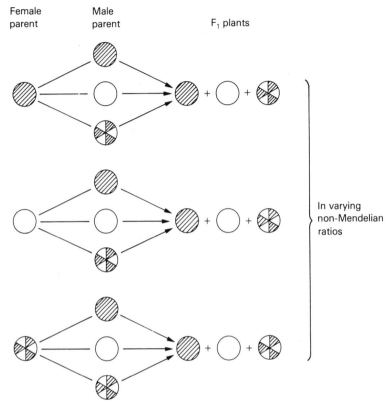

Female parent   Male parent   F$_1$ plants

In varying non-Mendelian ratios

Figure 4.11   Inheritance of defective plastids in *Pelargonium*. Shaded, open and sectored circles represent green, white or variegated plants (or regions of plants), respectively

mitochondrial DNA and this has been rigorously demonstrated by restriction enzyme digestion and fragment analysis of the mitochondrial DNAs of male fertile and male sterile strains of maize. However, in *Nicotiana* species the trait seems to be determined by chloroplast DNA and could be caused by a mutation in the gene coding for the ribulose bisphosphate carboxylase large subunit.

Of particular interest in the study of chloroplast genetics has been the unicellular alga *Chlamydomonas reinhardtii* as this has been very extensively studied. The organism is generally haploid and propagates itself either autotrophically in the presence of light or heterotrophically in the absence of light by mitotic cell division. The organism has a sexual cycle (figure 4.12) in which two individual cells morphologically identical but of opposite mating-type fuse together. (The two mating-types are referred to as mt$^+$ and mt$^-$.) Following cell fusion, fusion of nuclei and the two single chloroplasts donated by each strain takes place. On germination of the zygote, meiosis occurs giving rise to haploid

144

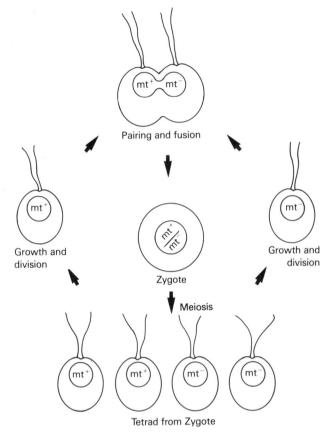

Figure 4.12   Life cycle of *Chlamydomonas reinhardtii*

progeny in which there is a 2:2 segregation of any characteristics determined by single chromosomal genes with contrasting alleles donated by the individual parents (including the mating-type characteristic). A number of characteristics have been shown to depart from this mode of inheritance however. The first such characteristic which was reported to exhibit non-Mendelian inheritance was a mutation leading to streptomycin resistance (sm-2) discovered by Sager. She found that the genotypes of progeny from a cross between sm-2-resistant and sm-2-sensitive strains depended on which mating-type carried which allele. If the mt$^+$ parent was sm-2-resistant then virtually all the progeny were also sm-2-resistant. This is an example of uniparental inheritance. A small proportion of zygotes, however, seem to be heterozygous and can give rise in subsequent mitotic divisions to either sm-2-sensitive or sm-2-resistant individuals. The incidence of such zygotes exhibiting biparental inheritance can be raised to 100 % by exposure of mt$^+$ cells to ultraviolet light irradiation immediately prior to mating. A range of other types of mutants showing a non-Mendelian inheritance pattern have also been isolated. Some of these are believed to be associated with mitochondrial DNA. Others, mostly induced by treatment with streptomycin, a fairly specific mutagen, have been ascribed to ctDNA. These include mutants which will only grow heterotrophically and mutants resistant to the various inhibitors of chloroplast protein synthesis (see section 4.4.2). These mutations have been mapped using the heterozygotes formed in the biparental mode of inheritance observed following mt^+-irradiation by ultraviolet light. Three types of segregation have been distinguished occurring in the mitotic divisions following such zygote formation. The first type of segregation (type I) is prevalent and preserves the heterozygous conditions in progeny. Type II segregation is the result of a non-reciprocal exchange event thought to be analagous to gene conversion and type III segregation is the result of a reciprocal exchange event (table 4.2). It is suggested that type III

Table 4.2   Segregation patterns during post-meiotic division of heteroplasmic
Chlamydomonas reinhardtii

	Zygote genotype	Genotypes of post-meiotic segregants
Type I (no segregation)	sm2-r/sm2-s	sm2-r/sm2-s + sm2-r/sm2-s
Type II (gene conversion)	sm2-r/sm2-s	sm2-r/sm2-s + sm2-r/sm2-r or sm2-s/sm2-s + sm2-r/sm2-s
Type III (reciprocal recombination)	sm2-r/sm2-s	sm2-r/sm2-r + sm2-s/sm2-s

segregation occurs when there is a crossover event between the gene in question and some hypothetical 'attachment point' (figure 4.13). The incidence of such type III segregants gives a measure of the map distance between the gene in question and the attachment point and can be used to construct a gene map. Similarly, re-combination levels between linked chloroplast genes can also be used to assess map distances. A map of the chloroplast genome as established by these methods is shown in figure 4.14.

The mechanisms of uniparental (maternal) inheritance are not yet clearly established. It had been assumed in early studies that differential genetic contributions of male and female to chloroplast inheritance resulted from non-inclusion of the chloroplast genome in the male gamete. However, in some cases (including a number of ferns) it seems that organelles including pro-plastids from the male gamete do not enter the egg cell, while in other cases (some algae) paternal organelles appear to be destroyed shortly after zygote formation. However, in some cases, including *Chlamydomonas* and a number of higher plants, zygotes are known to receive intact organelles from both parents. It has been suggested that ctDNAs from mt^+ and mt^- strains are differentially marked (possibly by methylation of the mt^+ ctDNA) prior to

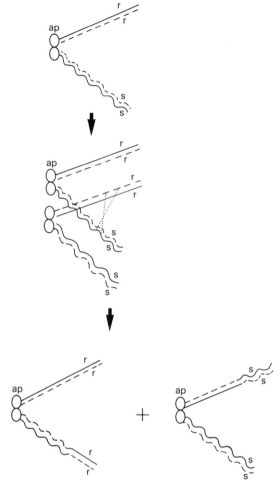

Figure 4.13   Mechanism of type III segregation in *Chlamydomonas reinhardtii*: r and s refer to resistant and sensitive alleles of a gene specifying sensitivity to a particular antibiotic (e.g., streptomycin sensitivity). Wavy and straight lines indicate DNA strands from two parents. Solid and broken lines of similar configuration are complementary DNA strands. Only previously synthesized DNA strands are attached to a hypothetical membrane attachment point ap which determines direction of movement at cell division

fusion, and that a gene closely linked to the mating type gene specifies a restriction enzyme which will degrade non-modified ctDNA from the mt⁻ parent. A gene closely linked to the mating-type locus has certainly been observed to influence the ratio of uniparental/biparental inheritance and this could be a manifestation of the above effect. Biochemical studies on the relative turnover rates of chloroplast DNAs from mt⁺ and mt⁻ strains following fusion have given ambiguous results, but it seems likely that differential breakdown of the two genomes may be important.

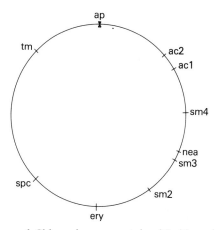

Figure 4.14   Gene map of *Chlamydomonas reinhardtii* chloroplast DNA. ap: attachment point, ac1 and ac2: acetate requiring genes, sm2, 3 and 4: streptomycin sensitivity genes, nea: neamine sensitivity; ery: erythromycin sensitivity, spc: spectinomycin sensitivity, tm: temperature sensitivity

4.5.2 Physical genetic mapping

In recent years mapping of chloroplast DNA has centred around the physical mapping of ctDNAs using restriction enzymes and molecular hybridization techniques. The first step is to produce a restriction enzyme map using an enzyme that will break the ctDNA into a small number of fragments and then sizing these fragments by agarose gel electrophoresis in the presence of standard DNAs of known size. Ordering of the fragments can be carried out by a number of different methods, the simplest of which involves partial digestion by the same restriction enzyme and examining the agarose gel electrophoretogram of the partial digest for larger DNA pieces whose length approximates to the total lengths of two or more fragments from the original total digests. These fragments must consequently lie next to one another. Physical maps have been constructed in this manner for *Chlamydomonas, Euglena*, maize, pea and spinach ctDNAs.

DNA/RNA hybridization can be used to determine the physical location of genes on the various restriction fragments of the ctDNA molecule. This can be done for ribosomal and transfer RNAs and any protein gene product whose messenger RNA can be isolated. The two copies of each of the rRNA genes from maize, spinach and *Chlamydomonas* ctDNAs are shown to be located in the inverted repeat segment mentioned previously (section 4.3.1) while *Euglena* has three copies of the rRNA genes arranged in tandem. The 5S rRNA gene is usually close to the 23S rRNA gene which is separated from the 16S rRNA gene by a spacer region of variable length. The 4.5S rRNA gene is also located close to the other rRNA genes – in spinach and maize probably between the 5S and 23S genes. The *Chlamydomonas* 23S RNA is thought to contain an intervening sequence which is spliced from the molecule following transcription.

tRNA genes have also been mapped onto ctDNAs from a number of species. In *Chlamydomonas* and *Euglena* these map all over the genome. Spinach has been most extensively studied. Chloroplast tRNAs from spinach have been resolved into 35 species, many of which have been assigned by hybridization with restriction fragments to particular regions. A small number of these are present in duplicate lying on the inverted repeat region between the rRNA cistrons the larger, non-repeated sector of the DNA. The bulk lie somewhere towards the middle of this non-repeated sector. None have been located in the smaller non-repeated segment.

The location of the gene for the ribulose bisphosphate carboxylase large subunit has also been determined in *Zea mays* and *Chlamydomonas reinhardtii*. In maize a 4 kilobase pair cloned fragment of ctDNA, which was capable of specifying the synthesis of this protein, was found to be located 30 kilobase pairs away from the 5′ end of the closest of the two sets of rRNA genes and 71 kilobase pairs away from the other. A similar approach in *Chlamydomonas* also located the gene on a fragment outside the inverted repeat and well away from the rRNA genes. The gene for the 32 000 molecular weight protein has also been mapped on maize ctDNA in a non-repeated region very close indeed to one of the repeated regions. Partial chloroplast gene physical maps for *Chlamydomonas*, maize and spinach are shown in figure 4.15.

If this and preceding sections have created the impression that the nucleus of plant cells plays little part in the specification of chloroplast characteristics nothing could be farther from the truth. A very large range of nuclear genes play a role in chloroplast development and maintenance. Nuclear genes have been implicated in barley, for example, in formation of the prolamellar body, its dispersion into lamellae and formation of grana. Nuclear genes control porphyrin synthesis and subsequent transformation of porphyrins into chlorophylls and cytochromes. They also control the synthesis of chloroplast lipids, including the phospholipids forming the bilayer, which is the fundamental basis of chloroplast membrane structure, as well as chloroplast DNA and RNA polymerases, most of the chloroplast ribosomal proteins and the soluble enzymes of the stroma. In *Chlamydomonas* a large number of mutants have been isolated, requiring acetate for growth, having lost photosynthetic function. These mutations have been located on virtually all of the 16 linkage groups of the alga. Clearly the nucleus plays a paramount role in chloroplast structure and functioning. The intensity of interest in the role of the chloroplast genome and its associated RNA and protein synthesizing systems probably arises from an interest in the question of why decentralization of information processing has occurred in eukaryotes and also in the possibility of complete dissection of this relatively small genomic system. In addition, the question of the evolutionary relationships between eukaryotic organelles and prokaryotic cells has undoubtedly been a major factor spurring research in this area. This is a question we will ponder in the final section.

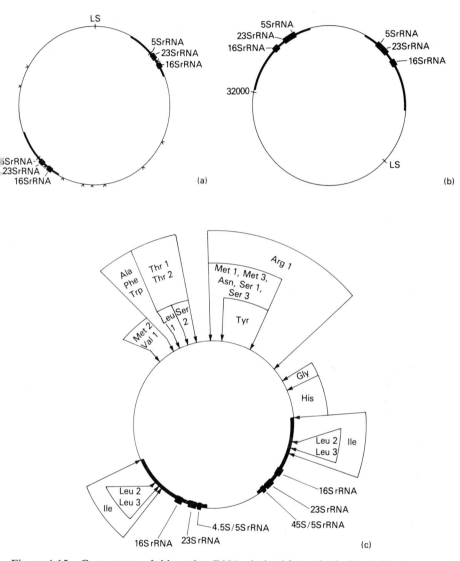

Figure 4.15   Genes maps of chloroplast DNAs derived from physical mapping.
(a) *Chlamydomonas reinhardtii*: arrows indicate approximate locations of tRNA genes.
LS: ribulose bisphosphate carboxylase large subunit (b) *Zea mays*: LS: ribulose
biphosphate carboxylase large subunit, 32 000: the 32 000 molecular weight membrane
protein (c) *Spinacea oleracea*: approximate regions of coding for various tRNAs are
indicated by amino acid symbols. In all cases thicker segments of the circular molecules
indicate the inverted repeat regions

4.6 Evolution of chloroplasts

The evolutionary relationships between the various classes of photosynthetic organisms is a topic which has generated much speculation and debate. On the basis of general structural organization living organisms are classified into two major groupings: prokaryotes and eukaryotes. The differences between the two groupings of organisms were discussed in section 1.1 and are summarized in table 4.3. According to the criteria listed here three types of photosynthetic organisms can be classified as prokaryotic. These are the classical photosynthetic bacteria now all grouped in the Rhodospirillales, the cyanobacteria

Table 4.3   A comparison of the properties of eukaryotic and prokaryotic cells

Feature	Prokaryotes	Eukaryotes
Size	Usually $1-10\ \mu$m diameter	Usually $10-100\ \mu$m
Nucleus	Only nucleoid-DNA-containing region of cell	True nucleus bounded by nuclear membrane
DNA	Simple duplex, not associated with histones Single linkage group	DNA in combination with histones Several chromosomes
Cell division	Binary fission	Mitosis (or meiosis)
Internal membranes	If present, very simple	Complex system of internal membranes
Ribosomes	70S	80S
Photosynthesis	Simple chromatophores	Chloroplasts
Oxidative respiration	Localized on cell membrane	Mitochondria

(formerly blue-green algae) and *Prochloron* the only known member of the newly recognized Prochlorophyta. The various types of algae and all higher green plants are considered to be eukaryotes. When we examine aspects of the photosynthetic process in these different groupings of organisms it is clear that the cyanobacteria and *Prochloron* show considerable similarities to green plants. In particular the use of water as a photosynthetic reductant with its conversion to oxygen (oxygenic photosynthesis), separates these classes of photosynthetic prokaryote from the Rhodospirillales (which have an anoxygenic photosynthesis). As already mentioned, cyanobacteria, *Prochloron* and photosynthetic eukaryotes have two photosynthetic systems (photosystem 2 and photosystem 1) which operate in series. This contrasts with the Rhodospirillales which lack photosystem 2.

These observations have led to the suggestion that, although the cyanobacteria and *Prochloron* stand on the same side of the prokaryotic/eukaryotic evolutionary divide as other photosynthetic bacteria, there must be a close evolutionary relationship between cyanobacteria and *Prochloron* on the one hand and eukaryotic chloroplasts on the other. All of the oxygenic photosynthesizers contain chlorophyll *a* as the primary photopigment but there are significant differences in the secondary pigments. Prochlorophyta or an evolutionary ancestor of this group has been implicated as the evolutionary precursor of green algal chloroplasts and subsequently the chloroplasts of all higher plants. This is because *Prochloron*, like all of these, uses chlorophyll *b* as a secondary photopigment. The cyanobacteria have phycobiliproteins as secondary pigments, however, and lack chlorophyll *b*. These features are shared with the Rhodophyta or eukaryotic red algae and this has been used to suggest an evolutionary link in the case of these two groups. Other groups of eukaryotic algae, e.g., chromophyta, which use chlorophyll *c* as a secondary pigment, are thought to have chloroplasts that have evolved from another type of prokaryotic cell, at present either extinct or undiscovered.

It is the precise nature of these evolutionary relationships which has been the topic of much recent discussion. The greater complexity and sophistication of the eukaryotic cells implies that the latter must have evolved from the simpler prokaryotes. The fossil evidence supports this sequence. The first eukaryotic fossils appeared about 10^9 year ago towards the beginning of the Cambrian period. Prokaryotic fossils provide evidence for the existence of prokaryotes as long as 4×10^9 years ago, well back into the Pre-Cambrian period. Therefore, with a fair degree of confidence, we can make the hypothesis that precursors of present-day oxygenic photosynthetic prokaryotes were involved in the evolution of photosynthetic eukaryotic plants.

There are two major theories for the evolutionary events which occurred in the transition from prokaryotic state to the eukaryotic state. The first of these has been referred to as the endosymbiotic theory. This hypothesis suggests that chloroplasts have evolved from primary events whereby host cells were invaded by precursors of present-day cyanobacteria or prochlorophyta. This established the elements of a symbiotic relationship so successful that it has survived in one form or another through 1000 million years. In the first stages of evolution of this relationship it is likely that the association was a fairly loose one, depending on the physiological advantages to the individual partners (photosynthetic products of the endosymbiont being exchanged for the relative protection of living in the cytoplasm of a larger and possibly more mobile organism), to gear together their growth rates. This might have been followed in the evolutionary timescale by the development of a more formal relationship between the partners in the endosymbiosis wherein their cell division and DNA replications might be tightly coupled by an interplay of regulatory products. A successful endosymbiotic partnership might also be expected to result in loss of some functions of members of the partnership when these were present in duplicate. If chloroplasts did develop from endosymbiotic photosynthetic prokaryotes, it is these, rather than the host cell, which suffered greatest loss of function. Isolated chloroplasts do have a wide range of metabolic activities and may even be capable of limited division, but no significant growth ever occurs.

Clearly, they have not retained the independence of their putative prokaryotic ancestor. The range of chloroplast components specified by nuclear DNA also attests to the deficiencies of chloroplasts compared with independent pro-karyotic cells. On the other hand, the host cell seems to have made little sacrifice. *Euglena* mutants lacking chloroplast DNA have been isolated which will grow heterotrophically and so clearly in this case, no vital function has been handed over to the chloroplast genome. By contrast, no such chloroplast DNA-less mutants have been isolated in *Chlamydomonas* or higher plants and so in these situations possibly ctDNA is coding for some function vital to the survival of the cell.

Proponents of the endosymbiotic theory for the origin of chloroplasts are in disagreement about the nature of the supposed host cell in the original endo-symbiotic event. Most consider that, as all eukaryotic organisms contain mitochondria, the acquisition of mitochondria by another endosymbiotic pro-cess involving entry and establishment in the cell of an aerobically respiring bacterium preceded the acquisition of chloroplasts. This might have led to the present diversity of autotrophic and heterotrophic eukaryotes. Other theorists believe that endosymbiotic evolution of chloroplasts preceded that of mitochondria and that this diversity arose by loss of chloroplasts from a species whose ecological niche favoured a heterotrophic mode of existence. The alternative hypothesis is that chloroplasts have developed from the progressive evolution of prokaryotes. The essential events here would be the encapsulation of photosynthetic membranes and a fragment (or whole molecule) of nucleoid DNA in the membranous vesicle. The encapsulated DNA would contain the genes for some of the photosynthetic protein components as well as other genes. Loss of those genes not concerned with chloroplast functioning would have trimmed the ctDNA to its present-day dimensions and functions. The idea that a whole molecule of cyanobacterial DNA would initially have been incorporated into the 'prechloroplast' is really more attractive as it would seem rather unlikely that a fragment of DNA would contain all of the rRNA- tRNA- and protein-specifying genes now known to be present on ctDNA, because of their scattered positions on the genome. Cyanobacteria, under particular culture conditions, are able to generate more than one genome per cell and consequently the above sequence of events is quite feasible. If a whole genome plus a portion of cell contents was included in a vesicle, this would really resemble a cell within a cell and might well be considered to be an internally generated endosymbiont.

Whichever of these two theories (if either) is the correct one there are remarkable similarities between cyanobacteria and chloroplasts over and above those already mentioned. These similarities have often been used by supporters of the endosymbiotic theory as evidence in favour of their ideas. However, many of these data are equally well explained by the progressive evolution theory. We will consider some of these similarities now and go on to evaluate them critically.

The organization of DNA in chloroplasts and prokaryotes, including blue-green bacteria, is very similar. In both cases there is a supercoiled, closed-circular structure, a feature which is considered to be primitive. The nuclear DNA of all eukaryotic cells is subdivided into a precise number of chromosomes (or pairs of chromosomes) characteristic for the species, each of which (or each pair of which) contains a distinct set of genes. In contrast, prokaryotes and chloroplasts possess just a single gene linkage group which may be present in more than one copy. The DNA of eukaryotes is complexed with histones, i.e., basic proteins which play an important role in maintaining chromosome structure and possibly play a part in control of gene expression. Such histones are not present in either prokaryotic or chloroplast DNAs. There are even similarities in arrangement of genes on the chromosomes of prokaryotes and chloroplasts. The mapping order of the genes for the various ribosomal RNA species is the same as that observed in *E. coli*, for example.

A study of the RNA species of chloroplasts and prokaryotes reveals further similarities. Extensive nucleotide sequence homologies have been discovered between the 16S ribosomal RNAs and chloroplasts, cyanobacteria and other bacteria, while no such homology has been discovered between the chloroplast 16S RNA and the equivalent cytoplasmic RNA (18S RNA) from eukaryotic cytoplasmic ribosomes. Chloroplast messenger RNAs lack the long polyA sequences attached to the 3′OH ends of eukaryotic cytoplasmic messengers. This is another prokaryotic feature. As already mentioned there are sequence similarities between cyanobacterial tRNAs and chloroplast tRNAs. The previously mentioned exchange of prokaryotic and chloroplast ribosomal subunits also demonstrates the close relationship between chloroplasts and prokaryotes. Detailed investigations on the structures and properties of a number of chloroplast and cyanobacterial proteins also support the idea of this close affinity. Particular similarities have been found in the phycobiliproteins (light-harvesting proteins) of red algae and cyanobacteria and ribulose bisphosphate carboxylase from cyanobacteria and higher plants.

In view of all of these remarkable similarities, it can be said with a considerable degree of conviction that there must be a close evolutionary relationship between chloroplasts and cyanobacteria. Can this be accepted as evidence for evolution of chloroplasts from an invading endosymbiont? It hardly can because an important feature of both the endosymbiotic theory and the progressive evolutionary theory (internally generated endosymbiont?) is that at a point in evolution a genome and the potentiality for processing that genome became sequestered within a membrane structure which isolated it from the rest of the cell. It is clear that in either case the essential components of the system have been conserved, possibly because of their relatively secure location, while the information storage and processing systems of the rest of the cell have been sophisticated in evolution possibly to meet the greater demands imposed by direct contact with the non-living environment. The divergence in properties of the chloroplast and extra-chloroplast nucleic acid and protein synthesizing systems would consequently be the result of differential rates of evolution in those two parts of the cell.

If these remarkable similarities between chloroplasts and prokaryotes fail to distinguish between the two major theories for chloroplast evolution, how then can the problem be solved? We can, of course, look to see whether we can find examples of equivalent processes to those postulated to have occurred in the evolution of chloroplasts taking place in present-day organisms, although we must be careful to point out that neither does their discovery prove a hypothesis nor the lack of discovery disprove it. Perhaps the most significant observations have been the existence of numerous endosymbiotic partnerships established between photosynthetic and non-photosynthetic organisms. A variety of these are described by Margulis in her excellent (although perhaps somewhat biased) book on the origin of eukaryotic cells. The most interesting is possibly the organism *Glaucocystis nostochinearum*, a photosynthetic alga which is clearly eukaryotic on the basis of the presence of a conventional nucleus, but with apparently blue-green chloroplasts. This organism is now considered to be derived from a green alga of the order Chlorococcales which has lost its green chloroplasts and has acquired endosymbiotic blue-green bacteria instead. There is also an example (*Cynophora paradoxa*) of a protozoan which harbours blue-green bacterial endosymbionts. There are also better known examples of green algae establishing stable endosymbiotic relationships with animal cells. The green hydra (*Hydra viridis*) is so coloured because it harbours the spherical green algae, zoochlorellae, in its endodermal cells. Another example which has been investigated in some detail is *Paramecium bursaria*. This paramecium is green in colour, and has endosymbiotic algae of the genus *Chlorella*. It has recently been observed that the optimal number of *Chlorella* cells per paramecium is regulated genetically by the host. The alga has been shown to enjoy what has been referred to as 'diplomatic immunity'. The paramecium will ingest and digest free-living *Chlorella* even though these might have no observable morphological differences from the immune endosymbiont.

These examples are very thought provoking and certainly show that endo-symbiotic associations have been established in many cases during evolution. It is not a big step to imagine the progressive loss of genetic capability of the photosynthetic partner until it has been reduced to the potentiality of present-day chloroplasts.

One problem with the endosymbiont theory is that it requires that transfer of some of the genes from the endosymbiont to the host nuclear DNA must occur. For example, the gene for the small subunit of ribulose bisphosphate carboxy-lase must have originated in the photosynthetic partner in the association. It is now located on the nuclear genome. The progressive evolution theory does not have this problem as initially both the 'nuclear' and 'protochloroplast' genomes could have been identical and random loss of genes from each genome could easily have led to a situation where a chloroplast protein could be composed of subunits coded by two genomes.

It is difficult to imagine how the full evolutionary history of the chloroplast might be established beyond doubt. The full answer may be lost in the biochemistry of organisms long extinct. It is nevertheless a most fascinating topic of discussion which will entertain biologists long after the problem of photosynthesis in all classes of photoautotrophic organisms have been solved.

Suggested further reading

Books

Beale, G. and Knowles, J. (1978). *Extranuclear Genetics*, Edward Arnold, London

Gillham, N. W. (1978). *Organelle Heredity*, Raven Press, New York

Margulis, L. (1970). *Origin of Eukaryotic Cells*, Yale University Press, New Haven

Sager, R. (1972). *Cytoplasmic Genes and Organelles*, Academic Press, London, New York

Tribe, M., Morgan, A. and Whittaker, P. (1981). *The Evolution of Eukaryotic Cells*, Studies in Biology, no. 131, Edwards Arnold, London

Growth and development of chloroplasts

Bradbeer, J. W. (1977). Chloroplasts, structure and development, in H. Smith (Ed.), *The Molecular Biology of Plant Cells*, Blackwell Scientific, Oxford, p. 64.

Green, P. B. (1964). Cinematic observations on the growth and division of chloroplasts in *Nitella*, *Am. J. Bot.*, **51**, 334.

Ohad, I. (1975). Biogenesis of chloroplast membranes, in A. Tzagaloff (Ed.), *Membrane Biogenesis*, Plenum Press, New York, p. 279.

Possingham, J. V. (1980). Plastid replication and development in the life cycle of higher plants. *Ann. Rev. Plant Physiol.*, **31**, 113.

Rebeiz, C. A., Smith, B. B., Mattheis, J. R., Cohen, C. E. and McCarthy, S. A. (1978). Chlorophyll biosynthesis: the reactions between protoporphyrin IX and phototransformable protochlorophyll in higher plants, in G. Akoyunoglou and J. H. Argyroudi-Akoyunoglou (Eds.) *Chloroplast Development*, Elsevier/North Holland, Amsterdam, p. 59

Chloroplast nucleic acid

Bedbrook, J. R. and Kolodner, R. (1979). The structure of chloroplast DNA. *Ann. Rev. Plant Physiol.*, **30**, 593

Ellis, R. J. and Hartley, M. R. (1974). Nucleic acids of chloroplasts, *MTP International Review of Science*, **6**, 323

Tewari, K. K. (1971). Genetic autonomy of extranuclear organelles. *Ann. Rev. Plant Physiol.*, **22**, 141

Synthesis of chloroplast proteins

Bottomley, W., Higgins, T. J. V. and Whitfield, P. R. (1976). Differential recognition of chloroplast and cytoplasmic messenger RNA by 70S and 80S ribosomal systems. *FEBS Lett.*, **63**, 120

Bradbeer, J. W. (1973). The synthesis of chloroplast enzymes. In B. V. Milborrow (Ed.), *Biosynthesis and its control in plants*, Academic Press, London, p. 279

Chen, K., Johal, S. and Wildman, S. G. (1977). Phenotypic markers for chloroplast DNA genes in higher plants and their use in biochemical genetics, in J. H. Weil and L. Bogorad (Eds.), *Nucleic Acids and Protein Synthesis in Plants*, Plenum Press, New York, p. 183

156

Ellis, R. J. (1975). The synthesis of chloroplast membranes in *Pisum sativum*. In A. Tzagaloff (Ed.), *Membrane Biogenesis*, Plenum Press, New York, p. 247

Kung, S.-D. (1977). Expression of chloroplast genomes in higher plants. *Ann. Rev. Plant. Physiol.*, **28**, 401

Chloroplast genetics

Bedbrook, J. R. and Kolodner, R. (1979). The structure of chloroplast DNA, *Ann. Rev. Plant Physiol.*, **30**, 593

Birky, C. W. Jr. (1976). The inheritance of genes in mitochondria and chloroplasts. *Bioscience*, **26**, 26

Gillham, N. W. (1974). Genetic analysis of the chloroplast and mitochondrial genomes. *Ann. Rev. Genet.*, **8**, 347

Sager, R. (1977). Genetic analysis of chloroplast DNA in *Chlamydomonas. Adv. Genet.*, **19**, 287

Evolution of chloroplasts

Bogorad, L. (1975). Evolution of organelles and eukaryotic genomes *Science*, **188**, 891

Margulis, L. (1971). Symbiosis and evolution. *Sci. Amer.*, **255**, 48

Index